Physical Science

Physical Science

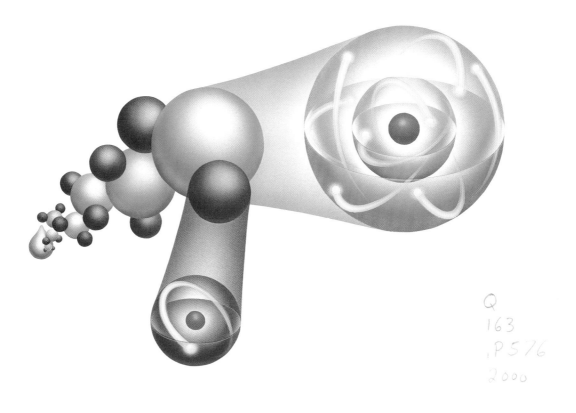

Time-Life Books   Alexandria, Virginia

Table of Contents

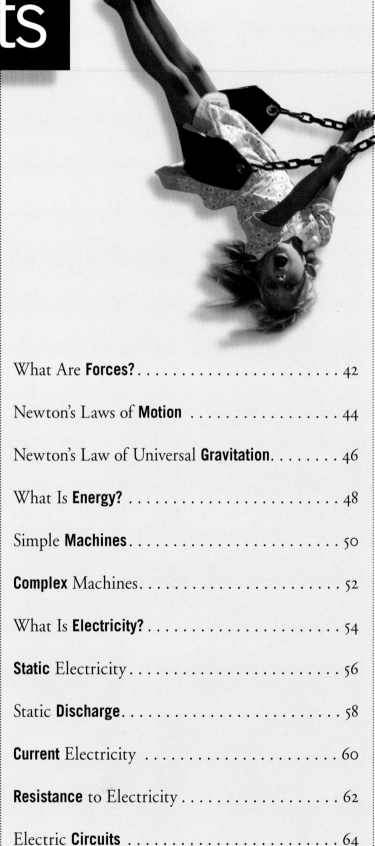

What Is Physical Science?

I n ancient times physical science was called natural philosophy, which meant thinking about and seeking to understand the natural world. Philosophers asked questions about the vastness of space and tried to determine the smallest part of matter.

Over the centuries people began to specialize. Astronomers investigated the stars and planets, physicists dealt with **gravity** and electromagnetic forces, and chemists delved into the structure of matter.

Today the physical sciences are still concerned with matter and **energy,** and the interactions in the fields of **mechanics, acoustics,** optics, heat, electricity, magnetism, **radiation,** atomic structure, and **nuclear** power. Scientists can explain what holds our universe together, what's inside an **atom,** and why people don't fall out of an upside-down roller coaster.

In a roller coaster, inertia *(page 44)* pulls the cars in a straight line, while gravity *(page 46)* pulls them down. A third force, called centrifugal (center-fleeing) reaction, pulls the riders upward, keeping them safely in their seats.

A fun house mirror is curved in several places *(page 87)*, reflecting a distorted image that is magnified in some parts and reduced in others.

Inside the Atom

The Tiny World of the Atom

What is matter made of? Do all physical objects—all things that take up space—have the same kind of basic building blocks? The Greek philosopher Democritus thought they did. About 430 BC he came up with the term **"atom"** to describe an unbreakable unit of matter. But it wasn't until the late 19th century that physicists began to prove the existence of atoms. They learned that atoms are the smallest units into which matter can be broken without releasing an **electric charge.** Atoms are also the smallest units of matter that have all the properties of a chemical element *(pages 10-11).*

Yet atoms are not solid. An atom has a tiny core, or **nucleus,** made of tinier particles known as **protons** and **neutrons.** The nucleus is surrounded by a whirling cloud of **electrons.** The attraction between the nucleus, which has a positive electric charge, and the electrons, which have negative charges, holds the atom together.

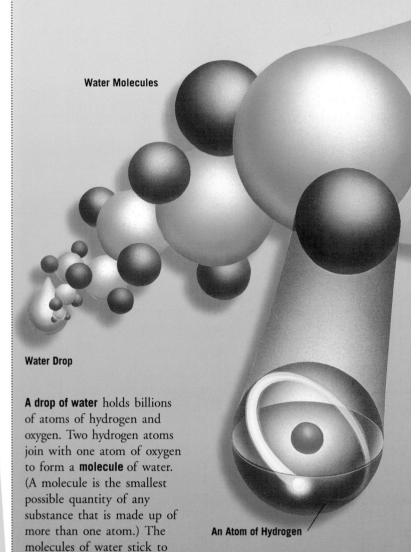

Water Molecules

Water Drop

A drop of water holds billions of atoms of hydrogen and oxygen. Two hydrogen atoms join with one atom of oxygen to form a **molecule** of water. (A molecule is the smallest possible quantity of any substance that is made up of more than one atom.) The molecules of water stick to one another to form droplets.

An Atom of Hydrogen

How Small?

How Small Is an Atom?

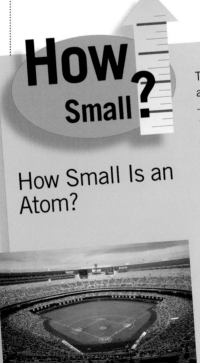

To get an idea of how small an atom really is, compare the Earth—which has a diameter of 11,617 km (7,926 mi.)—with a golf ball—which is 4.27 cm (1.68 in.) across. As small as the golf ball is, relative to the Earth, that's roughly how small an atom is compared with a golf ball.

And to understand how far the electrons of an atom are from the nucleus, imagine that the nucleus is the size of a grape sitting on the pitching mound of a ballpark. The electrons would be buzzing like fruit flies at the outer edges of the stadium. Most of an atom is empty space.

Hydrogen is the simplest possible atom. Its nucleus consists of a single proton *(orange)* and no neutrons. Only one electron orbits the nucleus. Hydrogen is by far the most common element in the universe, accounting for 87 percent of its **mass.** On Earth, hydrogen is present not only in water but also in all living plants and animals.

Electron

Nucleus

An Atom of Oxygen

Oxygen has a larger atom than hydrogen does. Its nucleus contains eight protons and eight neutrons. As in all atoms, its electrons are grouped into separate **orbits,** called "shells." Oxygen has two electrons in the first shell and six in the second. The six outer electrons combine with the two electrons from the two hydrogen atoms, joining the atoms into one water molecule.

Inside the Atom

An atom can be identified by its **subatomic particles**—protons, neutrons, and electrons. Its **atomic number** comes from the number of protons in its nucleus. Oxygen, with eight protons *(orange)* and eight neutrons *(green),* has an atomic number of 8. Atoms have the same number of protons and electrons, so it also has eight electrons. Protons have a positive electric **charge,** whereas neutrons have no charge. Electrons are negative. These tiny packets of energy are in constant motion around the nucleus. A certain number of electrons will fit into each shell around the nucleus. Oxygen has two electrons in its inner shell and six in the outer shell.

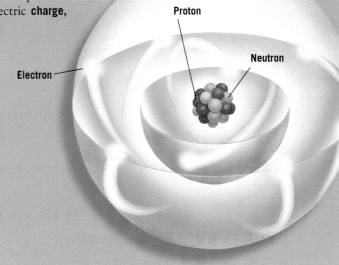

Proton

Neutron

Electron

Photographing Atoms

Atoms of xenon spell out a company name on a nickel surface in this 1989 photograph. Using a special device called a scanning tunneling microscope, a researcher at International Business Machines (IBM) in Almaden, California, was able to single out and move 35 atoms. Although xenon atoms are relatively large—for atoms —it would still take about 127 million of them to stretch across 2.5 cm (1 in.).

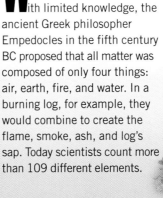

What's Matter?

With limited knowledge, the ancient Greek philosopher Empedocles in the fifth century BC proposed that all matter was composed of only four things: air, earth, fire, and water. In a burning log, for example, they would combine to create the flame, smoke, ash, and log's sap. Today scientists count more than 109 different elements.

The Elements

An **element** is a substance that cannot be broken down into any other substances. Hydrogen, for example, is an element. Water is not, because it can be broken down into hydrogen and oxygen. Elements like gold and lead have been known for thousands of years, whereas others, like radium, were not discovered until the 20th century. Some exist only for fleeting moments in particle accelerators.

In 1869, an insightful Russian chemist named Dmitry Mendeleyev *(far right)* came up with a way of showing all the elements known at that time. He invented a chart, now called the periodic table, that arranged elements into columns, called groups. The elements in each group shared similar chemical properties—they all reacted in the same proportions with oxygen, for example, or they all burned in water. Though it took some time for the idea to catch on—in part because nobody yet knew why different elements behaved similarly—the table is now a basic tool in the study of **chemistry.**

The Elemental Body

Like rocks, water, and air, you too are made of elements. As you can see in the chart at right, oxygen, carbon, hydrogen, and nitrogen occur in the largest amounts. These elements appear in almost all the body's tissues. Calcium and phosphorus are found mostly in bones and teeth. Your body even contains metals—magnesium, iron, copper, and zinc—as well as nonmetals like iodine and sulfur. Though present in only very small amounts, these elements are vital for good health.

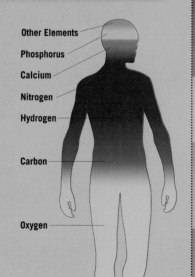

Other Elements
Phosphorus
Calcium
Nitrogen
Hydrogen
Carbon
Oxygen

The Periodic Table

Each element in the periodic table is defined by its **"atomic number"**—the number of **protons** its **nucleus** contains. With two protons, helium has an atomic number of 2, whereas lead's 82 protons give it an atomic number of 82. Columns in the table are known as groups, and horizontal rows are called periods. The periods all begin with a metal on the left and end with a gas on the right. (The elements in green, blue, and beige are nonmetals and include gases, liquids, and solids; all others are metals.) The symbols for most elements reflect their names in English, but some come from Latin—like Ag for silver, from the Latin word *argentum*.

1 H Hydrogen							
3 Li Lithium	4 Be Beryllium						
11 Na Sodium	12 Mg Magnesium						
19 K Potassium	20 Ca Calcium	21 Sc Scandium	22 Ti Titanium	23 V Vanadium	24 Cr Chromium	25 Mn Manganese
37 Rb Rubidium	38 Sr Strontium	39 Y Yttrium	40 Zr Zirconium	41 Nb Niobium	42 Mo Molybdenum	43 Tc Technetium
55 Cs Cesium	56 Ba Barium	57-71	72 Hf Hafnium	73 Ta Tantalum	74 W Tungsten	75 Re Rhenium
87 Fr Francium	88 Ra Radium	89-103	104 Unq Unnilquadium	105 Unp Unnilpentium	106 Unh Unnilhexium	107 Uns Unnilseptium

| 57 La Lanthanum | 58 Ce Cerium | 59 Pr Praseodymium | 60 Nd Neodymium | 61 Pm Promethium | 62 Sm Samarium |
| 89 Ac Actinium | 90 Th Thorium | 91 Pa Protactinium | 92 U Uranium | 93 Np Neptunium | 94 Pu Plutonium |

Dmitry Mendeleyev

The youngest of 17 children, Dmitry Mendeleyev (1834-1907) became a professor of chemistry in St. Petersburg, Russia. He came up with his first periodic table in 1869, and he later left gaps in it, convinced that new elements would be found to fit in the spaces. When the missing elements were discovered, they had exactly the properties he predicted.

					2 **He** Helium
5 **B** Boron	6 **C** Carbon	7 **N** Nitrogen	8 **O** Oxygen	9 **F** Fluorine	10 **Ne** Neon
13 **Al** Aluminum	14 **Si** Silicon	15 **P** Phosphorus	16 **S** Sulfur	17 **Cl** Chlorine	18 **Ar** Argon

26 **Fe** Iron	27 **Co** Cobalt	28 **Ni** Nickel	29 **Cu** Copper	30 **Zn** Zinc	31 **Ga** Gallium	32 **Ge** Germanium	33 **As** Arsenic	34 **Se** Selenium	35 **Br** Bromine	36 **Kr** Krypton
44 **Ru** Ruthenium	45 **Rh** Rhodium	46 **Pd** Palladium	47 **Ag** Silver	48 **Cd** Cadmium	49 **In** Indium	50 **Sn** Tin	51 **Sb** Antimony	52 **Te** Tellurium	53 **I** Iodine	54 **Xe** Xenon
76 **Os** Osmium	77 **Ir** Iridium	78 **Pt** Platinum	79 **Au** Gold	80 **Hg** Mercury	81 **Tl** Thallium	82 **Pb** Lead	83 **Bi** Bismuth	84 **Po** Polonium	85 **At** Astatine	86 **Rn** Radon
108 **Uno** Unniloctium	109 **Une** Unnilennium									

63 **Eu** Europium	64 **Gd** Gadolinium	65 **Tb** Terbium	66 **Dy** Dysprosium	67 **Ho** Holmium	68 **Er** Erbium	69 **Tm** Thulium	70 **Yb** Ytterbium	71 **Lu** Lutetium
95 **Am** Americium	96 **Cm** Curium	97 **Bk** Berkelium	98 **Cf** Californium	99 **Es** Einsteinium	100 **Fm** Fermium	101 **Md** Mendelevium	102 **No** Nobelium	103 **Lr** Lawrencium

Why the Table Works

The periodic table works because all the elements in a column have the same number of **electrons** in their outermost shells. And these electrons determine how an element will react with other elements. For example, lithium, sodium, and potassium are all in the first group and all have one electron in their outer shells. This is why they behave similarly.

Lithium

Sodium

Potassium

Elements and Bonding

Gold

The great family of known **elements** is still growing. At the end of the 20th century, scientists had named over 109 of them. Elements fall into three major categories. Most are metals, like lead, copper, and gold. Others are nonmetals, like neon, carbon, and sulfur. A few, like silicon, are called semi-metals, or metalloids, because they combine properties of both metals and nonmetals. Most elements are solids at room temperature, but a few are gases. Only two—mercury, a metal, and bromine, a nonmetal—are liquids.

Atoms bond, or link together, by sharing or swapping the **electrons** in their outer shells *(far right)*. If they didn't do this, the universe would be a huge mass of individual floating atoms. Elements form **bonds** in three different ways: by metallic, ionic, and covalent bonding, as seen opposite.

Known since ancient times, gold has long been prized for its beauty and rarity. Famous for its **malleability,** this metal can easily be shaped into thin wire or pounded into extremely thin sheets. A quantity of gold as small as 28 g (1 oz.), for example, can be stretched into a layer of foil with an area of 17 sq. m (187 sq. ft.). Gold's softness makes it ideal for jewelry, but to give their creations strength, jewelers must mix gold with other metals.

Mercury

Mercury is the only metal that is a liquid at room temperature, and it expands when heated. These properties make it useful in thermometers: The silver column of mercury rises. Some forms of mercury are very toxic, however, and can be dangerous even to touch.

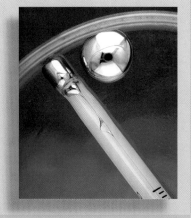

Neon

Neon and the rest of the elements along the right-hand column of the periodic table *(pages 10-11)* are known as "noble" gases. They almost never react with any other elements. Neon gas is often used in electric signs.

Potassium

Like the other metals in the first column of the periodic table *(pages 10-11)*, potassium does a surprising thing—it burns in water. This element is often used in fireworks, in which it creates a purple flame.

Carbon

Virtually all living cells hold carbon, the cornerstone of life on Earth. Pure carbon exists in two main forms: One is graphite, which is used in pencils. The other is diamond, which is among the hardest substances on the planet.

Covalent Bonds

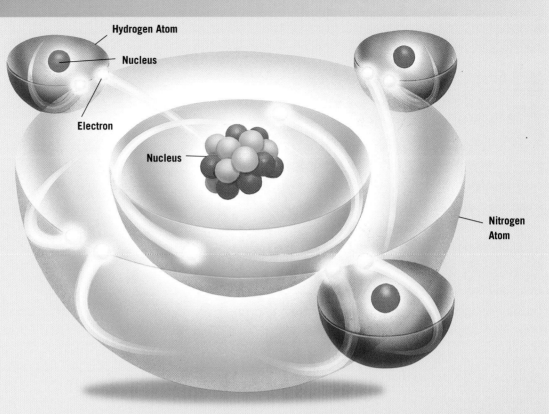

Some atoms join together by sharing electrons, a process called covalent bonding. A **molecule** of ammonia, for instance, is made of a nitrogen atom (N) bonded to three hydrogen atoms (H_3), as shown at right. Nitrogen, which has five electrons in its second shell, shares the three electrons from the hydrogen atoms to get a complete set of eight. Scientists use a shorthand description of the resulting ammonia molecule: They call it NH_3.

Hydrogen Atom

Nucleus

Electron

Nucleus

Nitrogen Atom

Metallic Bonds

Metals bond in an unusual way. Instead of sharing electrons between specific atoms, as in covalent bonds, they share their outer electrons with all the other atoms. These electrons can move around relatively freely, so metals tend to be good **conductors** of electricity.

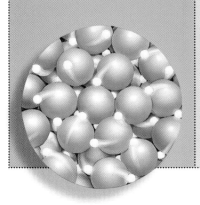

Ions and Ionic Bonds

Some atoms exchange electrons when they bond, rather than sharing them. Sodium, for instance, has just one electron in its outer shell. Chlorine needs one electron to fill its outer shell. Sodium will give up an electron and chlorine will take it. Short one electron, the sodium will become positively charged, and with an extra electron, the chlorine will become negatively charged. Charged atoms and molecules are called **ions**. When positive and negative ions come together, their opposite charges cause them to stick to each other through ionic bonding.

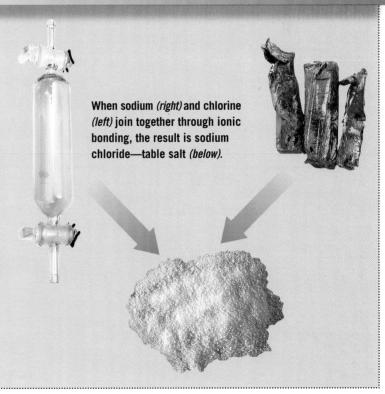

When sodium *(right)* and chlorine *(left)* join together through ionic bonding, the result is sodium chloride—table salt *(below).*

Nuclear Chemistry

I n the late 19th century, a French chemist named Antoine-Henri Becquerel discovered something strange happening in his lab. One of his chemical **compounds** was giving off mysterious rays that could develop photographic plates. He called these rays "radioactivity." Scientists later discovered that these rays were released when an atomic **nucleus** broke down into smaller nuclei.

Some **elements,** such as uranium and plutonium, are naturally radioactive, with their **atoms** decaying over time. Others can be made radioactive when they are bombarded with **neutrons** to split their nuclei. This is called **nuclear** fission, from the Latin word for "split," and it is the basis of nuclear weapons and nuclear **energy.**

Nuclear fusion is the opposite of fission. In fusion, atomic nuclei join together, or fuse, to form larger atoms. This happens in the cores of stars, where temperatures and **pressures** are high enough to overcome the **force** that holds **protons** and neutrons together inside an atom's nucleus. When nuclei fuse, they release large amounts of energy.

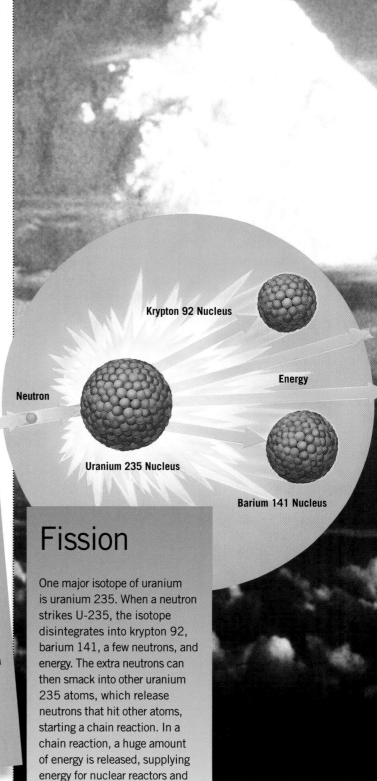

Krypton 92 Nucleus

Energy

Neutron

Uranium 235 Nucleus

Barium 141 Nucleus

Fission

One major isotope of uranium is uranium 235. When a neutron strikes U-235, the isotope disintegrates into krypton 92, barium 141, a few neutrons, and energy. The extra neutrons can then smack into other uranium 235 atoms, which release neutrons that hit other atoms, starting a chain reaction. In a chain reaction, a huge amount of energy is released, supplying energy for nuclear reactors and atomic bombs.

What Are Isotopes?

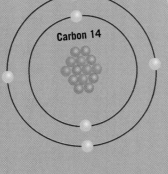

E ach atom of an element contains the same number of protons. Every carbon atom, for example, has six protons (orange).

Carbon 14

Carbon 12

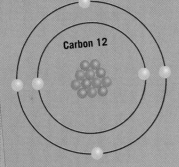

Most carbon atoms also have six neutrons (green), creating an atom of carbon 12 (left). But fewer than 1 percent of all carbon atoms have eight neutrons, resulting in carbon 14 (above). Atoms of the same element that have different numbers of neutrons are called isotopes.

Combining Nuclei

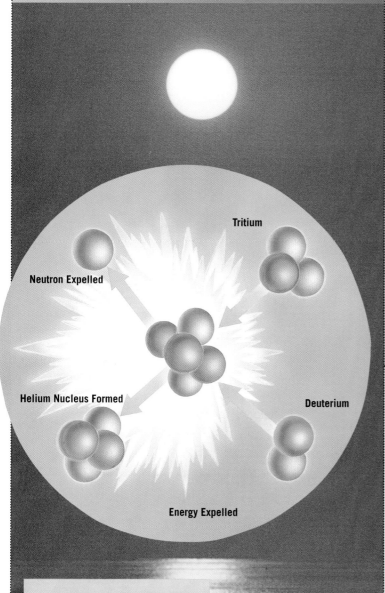

Tritium

Neutron Expelled

Helium Nucleus Formed

Deuterium

Energy Expelled

Fusion

The enormous amounts of energy that radiate from the Sun and other stars come from nuclear fusion. Under fantastic heat and pressure, isotopes of hydrogen —including deuterium, with one neutron, and tritium, with two—fuse together to form a helium nucleus with two protons and two neutrons. In the process, the isotopes release both energy and neutrons.

Radiocarbon Dating

Using an isotope of carbon, scientists can put a date on anything that was once alive —like the mummy below. Most of the carbon in living things is carbon 12, but a tiny fraction is the isotope carbon 14. Carbon 12 is stable, but C-14 is radioactive, decaying at a constant rate. New carbon 14 enters the body when we eat and breathe, so the ratio of carbon 14 to carbon 12 stays constant in a living body. When the body dies, it stops pulling in the carbon 14, and the isotope dwindles away. By measuring how little C-14 remains in an organism's tissue, scientists can estimate how long it has been since it died.

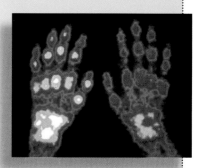

Radioactive Images

Doctors often use radioactivity to answer medical questions. X-ray images are made by sending **radiation** through a body onto film. Dense parts of the body, such as bones, absorb radiation better than soft tissues, such as muscles. So bones look more solid on

film. In the image below, left, the patient has swallowed a liquid containing barium, which blocks radiation well. This is why the digestive tract, holding the barium, shows up better than the rest of the image.

In another type of imaging, doctors inject the patient with radioactive material. The image then reveals where in the body the material is absorbed. In the hands above, the joints soaked up the radioactive "tracer," producing the bright spots. This shows a problem with the joints—in this case, arthritis.

Properties of Matter

Speaking very simply, matter can be defined as anything that takes up space. But that doesn't tell you very much about the matter. How big is it? Is it heavy or light? Is it cold or hot? How much does it weigh? Is it rough or smooth? What color is it? The answers to these questions will help you distinguish between one kind of matter and another. These are all properties of matter. Properties are the characteristics that describe and identify matter.

Imagine looking at an object through a pane of frosted glass. All you can see is a dark blob. What is it? You reach behind the glass to touch it—it's cold and wet. It has a smooth texture. You put some on your finger and taste it. It's very sweet. It's a scoop of ice cream! But you can't tell exactly what flavor. When you finally get to see it, you notice its color—white with a brown stripe running through it. Bingo! Now you know— it's a delicious scoop of chocolate ripple. You used the properties of ice cream—a kind of matter—to identify it.

Let's Compare

Physical and Chemical Properties

We use physical properties to describe matter. Color, shape, **mass**, weight, **density**, and temperature are all physical properties. The gooey texture of a melting marshmallow is another physical property. The appearance of the marshmallow changes, but what it is made of does not. Chemical properties, however, enable matter to change into new forms of matter. The flames leaping out of a fire are a result of chemical properties. Wood reacts with oxygen in the air and changes into something new.

Color

One physical property you use every day to distinguish between different kinds of matter is color. You can tell an apple from an orange, for example, just by looking at its color. Color is also an easy way to identify gemstones. Generally, rubies are red and sapphires are blue. Gemstones can be identified by other physical characteristics, too, such as hardness and the shape of their internal crystalline structure.

A Matter of Mass

You can't judge the weight of an object by its size. The three cylinders below are all different sizes but weigh exactly the same. Why? The size of an object depends on its **volume,** which is the amount of space it takes up. Weight, however, depends on mass *(page 42),* which is the amount of matter packed into that space. The lead cylinder *(left)* is the smallest, but it has the same mass as the wax and the wood, and it contains as much matter as the cylinders of wax *(center)* and wood *(right).*

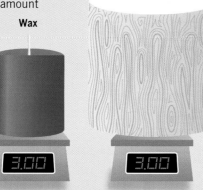

Wood

Wax

Lead

3.00 3.00 3.00

What's Density?

A room filled with people is densely populated. If only a few people are in the same room, the population is less dense. The same idea applies to matter. If a material has a lot of matter packed into a small space or volume, it is said to be very dense. If less matter is packed into the same space, that material has lower density. If you put a variety of liquids and solids into a jar, they will settle into layers according to density *(right).* The heaviest—or densest—ones will sink and the lighter—or less dense—ones will float. Which material is the most dense? Which is the least dense?

Cork

Oil

Plastic

Water

Grape

Syrup

Plasticity

If the shape of a material can be changed permanently by pressing or pulling, it has the physical property called plasticity. Clay is a plastic material. It can easily be pressed into different shapes. All metals have two particular kinds of plasticity, called **malleability** and ductility. Malleability means that a material can be flattened into a thin sheet. Ductility is the ability of a material to be pulled or drawn out into thin wires. Gold, for example, is both extremely malleable *(below)* and very ductile.

Viscosity

When you try to pour honey on a piece of toast, it seems to take forever for the liquid to drop from the spoon to the bread. On the other hand, when you pour water into a glass, it comes out so fast you can spill it all over the countertop if you're not careful. Both honey and water are liquids, but they move at different speeds. This **resistance** to flow is called viscosity. Thick, slow-moving liquids like honey and chocolate syrup have a high viscosity. By comparison, fast movers like water, alcohol, and baby oil have a low viscosity.

States of Matter

The world around us is made up of solids, liquids, and gases. When you sit in a chair, you're sitting on a solid. When you drink a glass of juice, you're swallowing a liquid. And when you take a breath, you're inhaling a gas. These basic forms of matter are called states of matter.

How are these three states of matter different? A solid has a definite shape and **volume.** A chair always looks like a chair and takes up the same amount of space wherever it's moved. A liquid also has a set volume, but it can change shape very easily. Milk poured from a rectangular glass will take on the shape of a round glass, but it will not fill the glass if the round glass is bigger. A gas has neither shape nor volume. If the air in a round balloon is released into a room, it will spread out until it fills the whole room.

Another state of matter is plasma *(below)*. This state is rare on Earth but is the most common in the universe.

What's Plasma?

Plasma is a gaslike mixture that exists primarily at very high temperatures, such as in the Sun and stars. At these high temperatures, **atoms** collide with one another so violently that their **electrons** are stripped away and move about independently. Plasma is not that common on Earth, but it does occur in flames, lightning bolts, and "plasma balls" *(right)*. In fluorescent lights, plasma is created by low pressure rather than high temperatures.

Different States

Each state of matter looks and acts the way it does because of the way its particles move. In a solid, particles are packed tightly together. They don't have enough kinetic **energy** to overcome their attraction to each other. Particles in a liquid can slide over each other, but they still don't have enough energy to completely overcome the **forces** that attract them. In a gas, particles finally have the energy to break free from each other and move away. This is known as the kinetic theory of matter.

Gas

Liquid

Solid

Would **You** *Believe?*

Liquid Glass

Did you know that glass has some liquid properties? Some people even call it a liquid solid. How can that be? The answer lies in the arrangement of the atoms that make up glass. Glass and sand are made of the same thing, silicon dioxide, but their atoms are arranged differently. Like most solids, sand *(top right)* has a rigid atomic structure. But glass particles *(bottom right)* are not so rigid, allowing them to move around each other. Over centuries glass can "flow." In

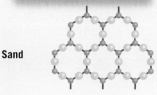

Sand

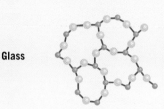

Glass

very old cathedrals and houses, you may notice that the glass in windows is actually thicker at the bottom than at the top!

Frozen Explosion

When water freezes, it expands, sometimes with a surprisingly strong force. In the experiment shown below, a scientist pours water into a cast-iron container with thick walls. Then he puts the container in a beaker of dry ice and alcohol, which produces freezing temperatures. As the water freezes, it expands, pressing against the walls of the container until it finally bursts through. The explosion is so fierce, pieces fly deep into a steel door 6 m (20 ft.) away!

Changes of State

Solid to Liquid

At 0ºC (32ºF), water changes from its frozen state—ice—to liquid. This is called a **phase change**. Luckily for the ice skater at right, the blade glides over only a thin layer of water, then the water freezes again.

Liquid to Gas

How does wet paint dry? How does a puddle disappear? In both cases it's due to evaporation. Bit by bit, water particles at the surface gain enough energy to change from a liquid to a gas and escape into the air.

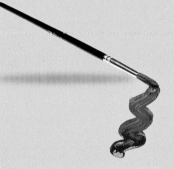

Gas to Liquid

Why do drops of water appear on the outside of a cold drink? The cold glass lowers the temperature of the air around it, removing energy from the water vapor in the air. As energy is lost, the vapor condenses—changes from a gas to a liquid—on the surface of the glass.

Sublimation

Changes of state don't always follow the solid to liquid to gas pattern. Sometimes a state is skipped. Dry ice *(right)* changes directly from a solid to a gas, skipping the liquid state. On very cold nights, water vapor in the air condenses onto car windows directly as ice, skipping the liquid state. This process is known as condensation.

What Is a Fluid?

When you stand on a windy beach and watch the waves crashing on the shore, you are seeing and feeling fluids in motion. Liquids, such as ocean water, and the gases that make up the air around you are both considered fluids. A fluid is any material that flows.

The characteristics of fluids explain how airplanes fly, how toothpaste comes out of a tube, and how small bugs can walk on water. Understanding how fluids behave also answers the basic question of why an object floats or sinks.

If you've ever tried to push an object under water, you've felt a **force** that seems to push back. This force, called buoyancy, is equal to the weight of the water pushed aside by the object. If the buoyant force is equal to the object's weight, the object will float. If not, the object will sink. The **density** (*page 17*) of the object and the amount of water it displaces—or pushes out of the way—influence the relationship between buoyancy and weight.

Floating

A boat can float for two reasons. First, the buoyant force of the water pushing up on the boat compensates for the boat's weight. Second, the boat's density is less than the water's. A 50,000-ton boat's hollowed-out shape full of air makes it less dense than even a small chunk of solid steel.

Walking on Water

Some insects and spiders like the one below depend on a little tension in their lives. A characteristic of fluids called surface tension forms an "elastic skin" on the surface of the water that's strong enough to support the bug's light weight. What causes surface tension? **Molecules** of water are attracted to each other. They pull at each other from all directions. At the surface, however, molecules only feel the pull from the sides and below since there are no water molecules above them. The effect of this downward pull is similar to that of a sheet stretched tightly across the surface of the water.

Sinking

An object sinks if its density is greater than the density of water and if it weighs more than the water's buoyant force. Even though this underwater robot is much smaller than the ship above it, it sinks because of its shape and density.

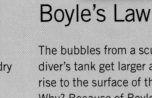

Charles's Law

If you put a balloon full of air near a container of very cold dry ice, the balloon will appear to deflate. That's because of Charles's law, which states that the **volume** of a gas shrinks with decreasing temperature.

Boyle's Law

The bubbles from a scuba diver's tank get larger as they rise to the surface of the water. Why? Because of Boyle's law. The water **pressure** decreases as the bubbles rise, so air molecules inside the bubbles can expand, spreading apart and increasing the size of the bubbles.

Diffusion

If a colored dye is added to a liquid, it will eventually spread out until it has mixed with the liquid. This is called diffusion. The dye diffuses because the particles of the liquid bump into the particles of the dye, pushing them around throughout the liquid (center and far right). Gas molecules also spread out, or diffuse, when released into the air. That's how the smell of baking cookies travels throughout a house.

Under Pressure

If you poke holes at different levels on the side of a plastic jug full of water, the water streams out at different rates. Near the top, the liquid will dribble out slowly. The closer you get to the bottom, the greater the **force** with which the water pours out. This happens because in fluids, pressure increases as you move downward.

Try it!

Pascal's Law

When you squeeze the bottom of a tube of toothpaste, why does toothpaste come out the other end? Because of Pascal's law: Pressure applied to a fluid travels unchanged throughout the fluid. When you squeeze the toothpaste tube, that pressure travels through the liquid, forcing it out through the opening.

Bernoulli's Principle

Take two empty aluminum cans and place them next to each other but slightly apart. Blow between the cans and watch what happens. Did the cans seem to magically move together? It's not magic, it's just Bernoulli's principle at work.

In the 1700s, the Swiss scientist Daniel Bernoulli discovered that the faster a fluid moves, the less pressure it exerts. When you blow between the cans, the air between them is moving faster than the air on the other side of each can. The slower-moving air on the outside exerts more pressure and pushes the cans together.

What Is Heat?

Scientists once thought heat was an invisible fluid called caloric. This fluid, they said, flowed into a substance when it was heated and flowed out when it was cooled. Today we know better. Heat is a type of kinetic **energy** *(pages 48-49)* that is released when **molecules** move. The faster the molecules move, the warmer a substance is.

Heat is always transferred in the same direction—from a warmer place to a cooler one. A glass of ice water feels cold to you because heat is drawn from your warm hand to the cool glass. The glass will feel cold until its temperature and the temperature of your hand are the same.

Heat travels in three ways. Conduction transmits heat through solids. Heat travels through liquids and gases by convection. Finally, **radiation** allows heat energy to travel through nothing at all—over great distances from the Sun or other stars.

Would You Believe?

Hot Iceberg

Even objects that are freezing cold have heat. There is more heat in an iceberg than in a cup of boiling water. How is that possible? It's because both the iceberg and the boiling water are made up of moving water molecules. Each molecule has energy. Because a big iceberg has many more water molecules than a small cup of boiling water, the iceberg has more heat energy.

Heating Things Up

If you've spent time at a pool, you've probably noticed that the metal handles of a pool ladder are sizzling hot long before the water in the pool is even warm. That's because it takes more of the Sun's energy to warm the water than it does to raise the temperature of the metal ladder. The ability of a material to absorb heat is a property called specific heat. Water has a high specific heat. A lot of energy is needed to raise the temperature of water even a small amount. Metal has a low specific heat. It doesn't take much energy to heat up metal.

Good Heat Conductors

A metal pot on a stove gets hot very quickly. But the plastic handle stays cool, because metal is a good heat **conductor,** and plastic is not. When a metal's molecules come in contact with a heat source, they begin to move faster and faster, crashing into one another at great speeds. This raises the temperature of the metal. One of the best heat conductors is copper, and that is why many pots have copper bottoms. The molecules in plastic are not easily set in motion by contact with heat. Plastic and other poor conductors, like wood and glass, prevent the transfer of heat.

Nature's Hidden Heat

Heat Released

Solid   Liquid   Gas

Heat Absorbed

Do you ever wonder how a tropical thunderstorm becomes a violent, spinning hurricane? A hidden form of energy, latent heat, helps explain it. This form of heat energy is locked in the water vapor of tropical storm clouds. When the water vapor changes into rain, energy is released and fuels the storm. Whenever a substance changes state—that is, into a gas, a liquid, or a solid—energy is either released or absorbed. In changes from gases to liquids to solids, latent heat is released. Changes in the opposite direction—from solids to liquids to gases—absorb heat. That's why steam, for example, can burn you.

How Heat Gets Around

Conduction

Conduction occurs when heat is transferred by direct contact. A cold spoon set in hot soup gets hot because the heat energy of the soup molecules moves to the cold spoon's molecules.

Convection

When a fluid heats up, like air above a hot pavement, it rises as it expands and gets lighter. When it cools, it gets heavy again and sinks. This movement creates convection currents.

Radiation

Invisible rays transfer heat from the Sun to sunbathing lizards. This process is called radiation. Any warm object gives off infrared radiation. When you stand in front of a fire, you feel radiant heat.

Try it!

Cut a piece of paper into a spiral. Tie a string to the center of the spiral, then hang it over a lighted lamp (left). The spiral will begin to spin. Why? The air warmed by the bulb rises, creating a convection current. This movement provides the energy that spins the spiral.

CAUTION!
Don't touch the light bulb with your hand or with the paper.

Measuring Temperature

What's the difference between heat and temperature? Heat is the total **energy** that an object has because its **molecules** are moving. Temperature is a measure of how fast those molecules are moving.

If you have a big pot and a small pot filled with boiling water, both pots will be the same temperature—100°C (212°F)—but the bigger pot will have more heat because it contains more moving molecules.

We measure temperature with a thermometer with scales that assign numbers to each temperature. The three most common scales used today are the Celsius, Fahrenheit, and Kelvin scales. Most of us measure temperature in degrees Celsius or degrees Fahrenheit. But scientists use a different scale, called Kelvin.

Early Thermometers

Italian scientist Galileo Galilei built the first thermometer in 1592. But it wasn't very accurate. Neither were the many other temperature measuring devices constructed over the next 125 years. But some were very unusual and attractive.

Florentine glassblowers made this beautiful thermometer *(below)* in order to help early meteorologists measure outdoor temperatures. The colored glass balls in each tube rise and fall in the water in response to changes in temperature.

In about 1657, another glassblower created a thermometer *(above, right)* to measure body temperature. A person blows on the ball at the bottom, which contains alcohol. The hot breath heats the alcohol, causing it to expand and rise up into the glass spiral. Small black glass beads mark the temperature scale. The height to which the alcohol rises within the spiral determines the person's temperature.

Temperature Scales

Below you can see how some familiar temperatures compare on the Celsius and Fahrenheit scales. The Kelvin scale *(right)* is used more often in science. The lowest Kelvin temperature is absolute zero, the temperature at which all molecules stop moving. No one has ever reached that cold a temperature, even in the laboratory.

Let's Compare

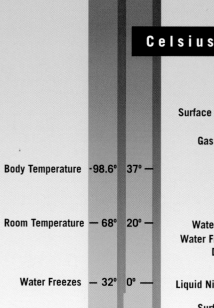

Kelvin

- −100,000,000
- Sun's Core — 15,000,000
- — 1,000,000
- — 100,000

Celsius

- — 10,000
- Surface of Sun — 6,000
- Gas Flame — 3,740
- Lava — 1,400
- — 1,000

Water Boils	373
Water Freezes	273
Dry Ice	194
	100
Liquid Nitrogen	77
Surface of Planet Pluto	43
	10
Helium Boils	4
Helium Freezes	1
Absolute Zero	0

	Fahrenheit	Celsius
Water Boils	212°	100°
Body Temperature	98.6°	37°
Room Temperature	68°	20°
Water Freezes	32°	0°
Mercury Freezes	-38.2°	-38.9°

Nature's Thermometer

If you don't happen to have a thermometer handy, nature can help tell the temperature outside. For example, crocuses open and close in response to temperature shifts. So sensitive are the flowers, in fact, they react to changes in temperature as small as 0.5°C (0.9°F).

People — Fahrenheit and Celsius

The next time you take your temperature, thank a German physicist named Daniel Gabriel Fahrenheit. In 1714, he developed the first reliable thermometer. Instead of using alcohol, he filled the closed tube with mercury, which expands more consistently and noticeably when heated. The other familiar temperature scale was developed in 1742 by a Swedish astronomer named Anders Celsius (right). His scale defined the freezing point of water as 100 degrees and the boiling point at 0 degrees. Sound backwards? It was only later that the scale was turned upside down, defining the freezing point as 0 and the boiling point as 100.

How Thermometers Work

Two of the most common thermometers are the liquid-in-glass thermometer (far right) and the digital thermometer.

A thermometer made of liquid in glass relies on the expansion of liquids—usually mercury or alcohol—to measure temperature. When the liquid gets warm, it expands and rises up a narrow tube. When it gets cooler, the liquid contracts and falls back down the tube.

In digital thermometers, changes in temperature affect the flow of electric current through metals inside the thermometer. Some metals resist the flow of electricity when heated. A tiny computer inside the thermometer translates the amount of **resistance** into a temperature value that appears as a number in the digital display window.

How Hot?

What happens when you stick a regular thermometer in flowing lava? It melts! So how do volcanologists (people who study volcanoes) take lava's temperature? They use a special kind of thermometer known as a thermocouple, which measures temperature by monitoring changes in an electric current. The probe that they insert in the lava is made of materials with very high melting points so it can withstand the hot temperatures.

Putting Heat to Work

E very time you start a car, refrigerate your food, or switch on the lights, you are putting heat to work. **Energy** from heat is changed into other kinds of energy to perform these tasks.

A car starts when a hot spark ignites fuel in the car's engine. The burning of the fuel releases chemical energy. The **force** of that **reaction,** triggered by heat, is turned into mechanical energy, which turns the wheels of the car.

Of course, heat energy doesn't always have to be converted into other kinds of energy to be useful. Many of the appliances in our homes rely on the movement of heat from one place to another. Refrigerators, for example, absorb heat inside and release it to the outside, keeping food cool so it doesn't spoil as rapidly.

Inside a car engine, heat energy is transformed into mechanical energy. The sequence of drawings at right shows how.

Engine

Engine Cycle

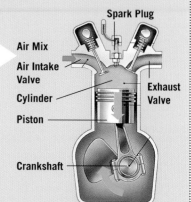

Spark Plug
Air Mix
Air Intake Valve
Cylinder
Piston
Exhaust Valve
Crankshaft

Injection

The air intake valve opens and the piston moves down, drawing a mixture of air and gas into the cylinder.

Compression

The piston moves back up and compresses, or squeezes, the mixture of air and gas in the cylinder, which heats it.

Combustion

A spark from the spark plug ignites the air/gas mixture and it explodes. The sudden expansion of gases from this explosion forces the piston down and makes the crankshaft go around. This burst of energy is transferred to the wheels of the car.

Exhaust

The piston rises again, pushing out the exhaust, or waste from the burned gasoline. Then the process begins again.

1

2

3

4

An energy chain is the process of one form of energy turning into another and then into another and so on. For example, many different kinds of energy come together to produce the lights and sounds of a carnival. The chain starts at a power plant. At a hydroelectric plant *(1),* for example, the mechanical energy of moving water turns huge turbines. These turbines rotate shafts inside generators *(2),* which transform the mechanical energy into electric energy. Power lines *(3)* deliver the electric energy to consumers, like the people at a carnival *(4).* There it is changed into several different kinds of energy, including light and sound.

Thermal Pollution

Changing heat into other kinds of energy isn't a perfect process. Some heat is always lost. When this lost heat is released into the environment, it increases air and water temperatures and can endanger plants and animals. This threat is called thermal pollution.

Trapping Heat

What do ducks and thermos bottles have in common? Both are designed to stop heat transfer. A duck's skin has a covering of soft, fluffy feathers, called down, that forms a kind of net to trap air. The down keeps a duck's body heat from escaping to the cool air around it. The design of a thermos bottle also prevents heat movement. Inside are two glass walls.

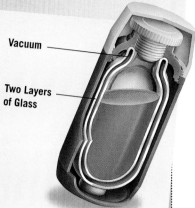

Vacuum

Two Layers of Glass

Between these walls is a vacuum, a space that contains no matter. Heat cannot move through a vacuum by convection because there is no matter to transfer the heat. In addition, the glass walls are silvered to reflect radiant heat waves *(page 23)* back into the bottle.

Keeping Cool

Evaporator

Condenser   Compressor

A refrigerator works by combining two basic physical principles: evaporation and condensation. When a liquid changes into a gas by evaporation, heat is absorbed. When a gas changes back into a liquid by condensation, heat is released. Inside the refrigerator, a special liquid called a refrigerant is pumped through a series of tubes. As the liquid passes through the evaporator in the freezer compartment, it changes into a gas and absorbs heat from inside the refrigerator. In the condenser, the gas is changed back into a liquid, releasing its stored heat to the outside. Then the liquid refrigerant flows to the evaporator again, starting the cycle anew.

Solutions and Other Mixtures

Ocean water. Sweetened coffee. Tea. These are all examples of solutions. A solution is a special kind of mixture that is formed when one substance dissolves into another. The particles of each substance are spread out evenly and are so small that they can't even be seen with a microscope.

Not all mixtures are solutions. A mixture is simply the combination of two or more substances. A tossed salad, for example, is a mixture of lettuce and other vegetables. Unlike a solution, in this kind of mixture you can easily distinguish the different parts and those parts aren't evenly distributed.

In mixtures, the substances are physically combined but not chemically combined. Because of this, it is pretty easy to separate the substances that were brought together to make the mixture.

The air around us is an example of a mixture called a colloid. In a colloid, particles are evenly spread out, as in a solution, but they are much bigger. Sometimes you can see the particles dispersed through air when sun rays pass through it *(above)*. A colloid can consist of a gas in a solid (Styrofoam), a liquid in a liquid (milk), or a solid in a gas (smoke).

Mixtures for Breakfast

Different kinds of mixtures surround us every day. Many of the foods we eat and drink at breakfast, for example, are mixtures. Even the utensils we use may be made of materials that have been mixed together.

Some mixtures, like solutions, have unique characteristics. Scientists have given special names to other kinds of mixtures that share special traits. Alloys, suspensions, and colloids are examples of such special mixtures.

Alloy
A stainless steel spoon is an alloy, a solution made from a mixture of the metals chromium and iron.

Suspension
Orange juice is a suspension. It's a liquid in which visible particles—the pulp—eventually settle to the bottom.

Colloid
Milk is a mixture of microscopic particles of different sizes called a colloid.

Solution
A cup of tea is a solution of tea (from the tea leaves) and sugar spread out evenly throughout warm water.

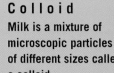

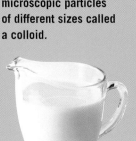

Hetero-geneous Mixture
Cereal with fruit and milk is a heterogeneous mixture. You can easily pick out the different parts in it.

Separating into Parts

Filtering Mixtures

Usually mixtures can easily be separated into their parts. The method used depends upon the characteristics of the mixture. If the particles in the mixture are different sizes, they can be separated by filtration. In a coffee maker, hot water runs over the coffee grounds, dissolving the coffee in the liquid. The liquid passes through the tiny holes in the filter, but the grounds, which are too large to fit through the holes, are left behind.

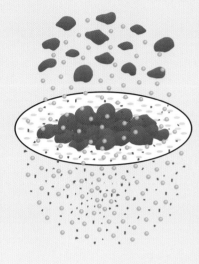

Spinning Mixtures

In some mixtures, like colloids, the particles are too tiny to separate with a filter. Instead, the particles may be separated according to their weight using a centrifuge. The centrifuge spins the mixture at high speed, causing denser particles to sink to the bottom while lighter ones rise to the top. Spinning blood, for example, separates the heavy blood cells from the lighter liquid plasma *(right)*.

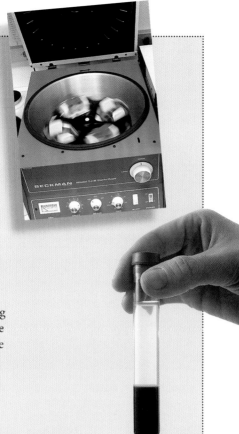

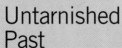

Would You Believe?

Untarnished Past

Before the 1600s, it wasn't necessary to polish silver, because it didn't tarnish. Silver tarnishes, or discolors, when it reacts with sulfur in the air. Before the 17th century, air contained very little sulfur. Over time, though, human activities, such as burning fossil **fuels**— coal, gasoline, or natural gas— increased the amount of sulfur in the air. Today sulfur dioxide is one of the most troublesome air pollutants as well as the reason that silver tarnishes so quickly.

Disappearing Act

When you make lemonade from a powder or put sugar in tea, you watch as solids dissolve in water. The solid particles appear to gently melt into the water, as in the

photograph at left, but there's actually a fierce attack going on. The drawing above shows that water **molecules** *(blue)* force their way into clusters of solid particles *(purple and pink)*, separating them from one another. The water molecules then surround the freed particles and prevent them from regrouping. So many substances can dissolve in water that water is known as the "universal **solvent**."

What Is a Compound?

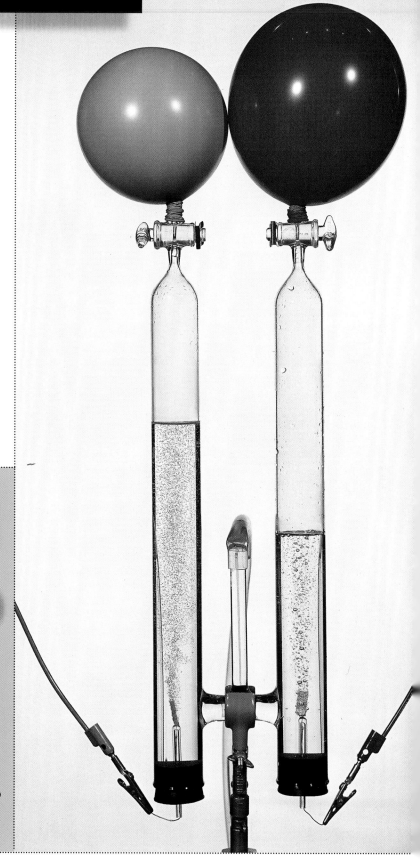

When two or more **elements** bond together chemically, they form a new substance with its own properties. This combination of elements is called a **compound.** Water is a compound made of hydrogen and oxygen. Carbon dioxide, which is what you release when you exhale, is a compound made of carbon and oxygen. Most substances around you are compounds or mixtures of compounds.

All compounds are made up of **molecules.** A molecule is the smallest part of a compound that has all the properties of that compound. For example, a molecule of water is made up of two **atoms** of hydrogen and one atom of oxygen. That molecule has the same boiling point, melting point, taste, color, and so on that a glass full of water molecules has. However, if you broke the molecule down into the atoms that form it, the hydrogen and oxygen would have their own unique properties that are entirely different from water.

Let's Compare

Compounds and Mixtures

Combining elements to form a compound is very different from just mixing elements together. In a mixture of iron filings and sulfur, for example, each element holds on to its original properties. You can separate the mixture pretty easily by dragging a magnet through it, attracting all the iron filings *(above)*. But if you heat the mixture, a **chemical reaction** takes place that bonds the elements together. Not only does the new compound *(left)* look different, but it also has no magnetic properties.

Separating Compounds

Unlike mixtures, which are fairly easy to separate, the strong **bonds** in a compound can be broken only by chemical means. Heat is one way to break the bonds holding elements in a compound together. Another method is with electricity.

When an electric **current** passes through two tubes full of water *(below)*, it breaks the tight bonds between the hydrogen and oxygen atoms in each water molecule. At left, oxygen atoms bubble up to fill the blue balloon, while hydrogen atoms fill up the red balloon. Because there are two hydrogen atoms for every atom of oxygen in a water molecule (H_2O), the red balloon fills up faster.

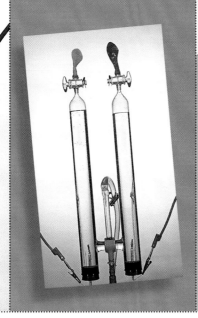

(pages 10-11)

Then & NOW!

Have you ever wondered why the periodic table of elements *(pages 10-11)* uses symbols, such as Au for gold, Hg for mercury, and Pb for lead? It's because in 1814 scientists began using alphabetical symbols for the elements based on their Latin names: aurum, hydrargyrum, and plumbum. Before that, the symbols were even more random—and artistic. Scientists often used the Sun as a symbol for gold, the caduceus—or staff—of the god Mercury for mercury, and the scythe—or curved blade—of the god Saturn for lead *(above)*.

	1500s	1600s	1700s	1783	1808	1814
Gold						Au
Mercury						Hg
Lead						Pb

What a Difference an Atom Makes

The two compounds at right may not look alike, but they are made of the same elements: iron and chlorine. The only difference between them is one atom. Ferrous chloride ($FeCl_2$) has one fewer atom of chlorine than ferric chloride ($FeCl_3$) does. The proportions of the elements must remain constant or else the compound changes into something else. Scientists refer to this as the law of constant composition.

Ferrous Chloride

Ferric Chloride

A Shorthand for Scientists

Scientists can refer to a compound in two ways—by its chemical name or by its chemical formula. Both are based on the elements in the compound. The chemical name for chalk *(right)* is calcium carbonate. But in writing, it's much faster for scientists to refer to chalk by its chemical formula: $CaCO_3$. The formula is a combination of the letters that represent its elements: calcium (Ca), carbon (C), and oxygen (O). The number "3" tells us that for every atom of calcium and carbon there are three atoms of oxygen.

$$CaCO_3$$

A Look at Chemical Reactions

id you know that as you read this page, you are surrounded by **chemical reactions?** They are constantly occurring in your body as food is turned into **energy.** The leaves on plants and trees are busily converting sunlight, carbon dioxide, and water into sugar and oxygen through the chemical reaction called photosynthesis. Planes flying overhead and cars driving by rely on chemical reactions to provide the energy to make them move. But what is a chemical reaction?

A chemical reaction happens when two or more substances react with one another. The reaction causes the substances to break apart and then recombine to form new substances. The original substances are called the reactants. The new substances are called the products. The products of a reaction usually have very different physical and chemical properties from the reactants.

Every reaction needs energy to break the **bonds** between the reactants. When new bonds are formed, energy is released. This energy may be in the form of heat, light, sound, or electricity.

Let's Compare

Heat Absorbers and Heat Releasers

Exothermic Reaction

When old bonds are broken and new ones are created in a chemical reaction, energy is both released and absorbed. If the reaction releases heat, like in a fire, it is called an exothermic reaction. If it absorbs heat, as do the cold packs athletes use on injuries, it's called an endothermic reaction.

Endothermic Reaction

Creating Something New

In every chemical reaction, from exploding fireworks to digestion, bonds are broken and new bonds are made. A simple example is what's going on when methane—the main part of natural gas—burns. Before the reaction (1), each methane **molecule** has four hydrogen **atoms** (red) bonded to one carbon atom (black). Pairs of oxygen atoms (blue) are bonded together in the surrounding air. When methane burns, it reacts with the oxygen molecules in the air and all the bonds break (2). The pieces recombine and form new bonds, making carbon dioxide and water (3).

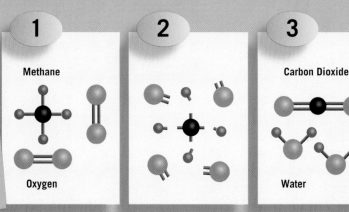

1 Methane / Oxygen

2

3 Carbon Dioxide / Water

Chemical Equations

If scientists always had to write in words what happens during a chemical reaction, many of them would suffer from hand cramps. To make their work go faster, they often use symbols instead of words.

The description looks like an equation. On one side are the reactants and on the other are the products. In between is an arrow to show which way the reaction is occurring. This arrow stands for "yields" rather than equals.

Below is the chemical equation for the formation of carbon dioxide: One carbon atom combined with two oxygen atoms yields carbon dioxide. Beneath it is a graphic representation of the reaction.

$$C + O_2 \rightarrow CO_2$$

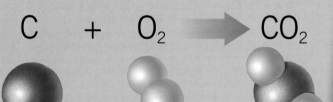

People — Antoine Lavoisier

Many people consider the French scientist Antoine-Laurent Lavoisier to be the father of modern chemistry. In 1789, he became the first scientist to demonstrate the truth of the law of conservation of matter. He was known for his meticulous approach to experiments, and he was assisted by his wife, Marie-Anne, a talented chemist herself. Marie-Anne documented her husband's work with detailed drawings.

A Rusty Reaction

When iron is exposed to air and moisture, it combines with oxygen, forming iron oxide, or rust. In this reaction, the iron gains oxygen while the air and water lose it.

Scientists call this process an oxidation-reduction reaction, or "redox" for short. In every redox reaction, one substance gains oxygen while another substance loses it.

Inside a Battery

Batteries hold chemical energy that can be changed into electric energy. Running down the center of a battery is a moist paste or liquid that is called an electrolyte. An electrolyte is a chemical, or a mixture of chemicals, that conducts electricity. When you put the battery in a flashlight or Walkman and turn it on, chemical reactions cause **electrons** to flow from the negative electrode through the device and back into the battery at the positive electrode. This flow is what we call electricity.

Positive Electrode

Electrolyte

Negative Electrode

Rates of Reactions

Have you ever wondered why food is put in a refrigerator? Some things taste better cold, but that's not the only reason to put them in a refrigerator. When food is cool, it doesn't spoil as quickly. Particles move more slowly at lower temperatures, so **reactions** happen at a slower rate. For a **chemical** reaction to happen, particles must collide with enough **energy** to break **bonds.** If they don't have the energy, they just bounce off each other.

Temperature is just one way to control the rate or speed of a chemical reaction. Other important factors are surface area, **concentration,** and catalysts. Catalysts are substances that alter the rate of a reaction without being changed by it. When the reaction has run its course, the catalyst can be recovered.

Fuel Cells

On a space mission, the spacecraft needs electricity to run and the astronauts need water to drink. A chemical reaction aided by a catalyst in the spacecraft's **fuel** cell provides both.

A fuel cell is a device similar to a battery. It converts chemical energy directly into electric energy. The fuel cell in a spacecraft uses a metal catalyst such as platinum to help combine hydrogen and oxygen. This chemical reaction produces electricity and also forms water.

The astronauts use the water for drinking, washing, and preparing dehydrated food.

Unlike a battery, a fuel cell doesn't need to be recharged. As long as there is a supply of hydrogen and oxygen, the fuel cell can provide as much energy as the craft needs to function.

Surface Area

Thin, slivered potatoes fry faster than large chunks. The speed of the reaction increases because more of the potato is exposed to the hot oil. Whenever one of the reactants in a chemical reaction is a solid, the reaction can be sped up if the solid is broken into smaller pieces. The greater the surface area, the faster the rate of reaction.

Concentration

When people ask for a strong cup of coffee, they want a solution with a lot of coffee suspended in it. This is called the solution's concentration. A highly concentrated solution *(near left)* speeds up a reaction because there are more particles of the reactants in it to interact with each other. A weak solution *(far left)* will cause a slower reaction rate because it contains fewer particles that can interact.

Temperature

In 1997, scientists discovered a 20,000-year-old woolly mammoth frozen in the permafrost of Siberia. The 3.4 m (11 ft.) creature was so well preserved that plants frozen in the soil around it were still green. Why hadn't the mammoth decayed? Because it was frozen.

Particles have less energy in cold temperatures, so they don't collide as often. Slower particles mean a slower rate of reaction. At higher temperatures, they have more energy and collide more often. This increases the reaction rate.

Scientists can learn a lot about these animals that have been extinct for 10,000 years.

A Gallery of Catalysts

Zeolites

A catalyst's main job is to bring **molecules** together so they can react. A family of catalysts called zeolites *(right)* do this particularly well. Their honeycomb shape gives them a large surface area on which reactions can happen. These catalysts are important in many industries, including the manufacture of gasoline.

Enzymes

A sliced apple turns brown because of another kind of catalyst, called an enzyme. An apple's enzymes react with oxygen and other molecules to quickly produce the brown pigment that covers the exposed flesh of the apple. This protective barrier slows down the rate at which oxygen reaches the interior of the apple, so the fruit doesn't rot as fast.

Deadly Catalyst

Not all catalysts are good guys. The hole in the ozone layer, the part of Earth's atmosphere that protects us from the ultraviolet radiation of the Sun, is caused by a catalyst. Chlorofluorocarbons (CFCs), chemicals long used as refrigerants and in aerosol sprays, decompose and release chlorine into the atmosphere. Chlorine acts as a catalyst in a reaction that turns ozone into oxygen. Chlorine is left unchanged by the reaction, so it continues to destroy ozone.

Chemical Detectives

Is the water supply poisoned? What's making a patient so sick? What killed Napoléon? These are all questions that **chemistry** can help answer. When investigators need to identify a substance or a virus, determine the **concentration** of a solution, or answer even more exciting "whodunit" questions, they often turn to chemists to fill in the blanks.

Chemists have a variety of techniques they can use to solve mysteries. In some cases, simply knowing the physical and **chemical** properties of a substance is enough to identify it. Finding out what's in a substance is called qualitative analysis. Determining how much of each ingredient is in the substance is known as quantitative analysis.

We rely on the expertise of chemists for everything from food safety to the well-being of the environment.

When you need to know if the water is safe to drink or the air is safe to breathe, call on an environmental chemist. These scientists run chemical tests on samples of water, air, and soil to monitor levels of **pollution.** They can determine what chemicals, if any, are in the environment and how concentrated these chemicals are in the water, air, or soil. They might check river water for levels of pollutants such as fertilizer, waste, and **acid.** In some cases they can add chemicals like fluorine or chlorine to counteract a dangerous problem.

Learning about the Past

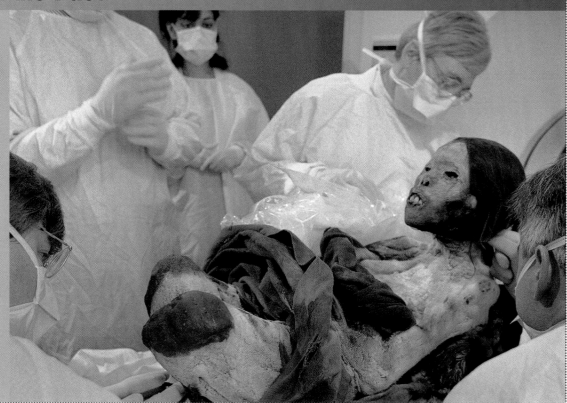

In 1995, archaeologists found the frozen body of an Inca girl who had been sacrificed on a Peruvian peak more than 500 years ago. Scientists used chemistry, among other things, to find out details about her life. Tissue samples from her stomach showed that she had eaten a meal of vegetables six to eight hours before she died. Samples from her knees showed that she had the healthy bones of a teenager. Scientists even hope to capture DNA from cells so they can trace her descendants and locate her living relatives. Who knows, one of them could be you!

Trial by Fire

Why do fireworks explode with such a brilliant display of color? The colors are created because different metals give different colors to a flame. This fact can also help scientists identify an unknown metal **compound.** For example, barium burns with a green flame, strontium burns with a red one, and sodium burns with a yellow flame *(below).*

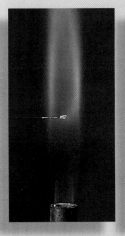
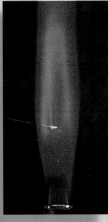

DNA Fingerprints

Did you know that your DNA, like your finger-prints, is unique to you? Detectives trying to solve a crime can use this fact to identify a suspect. Forensic scientists run a test called electrophoresis on a sample of hair, skin, or blood. This test separates the DNA, or genetic material, from the rest of the sample. Investigators then try to match the DNA profile with a suspect, just as they would with a regular fingerprint. That's why this process is sometimes called genetic fingerprinting.

Would You Believe?

Drop-Dead Decorating

When French emperor Napoléon Bonaparte died in 1821, doctors said the cause of death was a stomach ulcer that had turned cancerous. More than a hundred years later, scientists analyzed a lock of Napoléon's hair and found traces of arsenic, a deadly poison. Had he been murdered? Maybe. But maybe not.

Back in Napoléon's time, a green pigment, or coloring, used in fabrics and wallpapers contained the compound copper arsenite. If this paper became damp and moldy, the arsenic was released as a poisonous gas. A piece of wallpaper allegedly from the room where Napoléon died was tested, and scientists found that its green coloring did indeed contain arsenic. The concentration wouldn't have been enough to kill him, but it would have made him sicker. It could also explain why he had traces of arsenic in his hair.

Unfortunately, there's no way to know for certain how Napoléon died, or even to be sure the scrap of wallpaper really was from his room. But in an engraving that's considered an accurate portrayal of the scene around his deathbed, you can just see some wallpaper in the background *(right).*

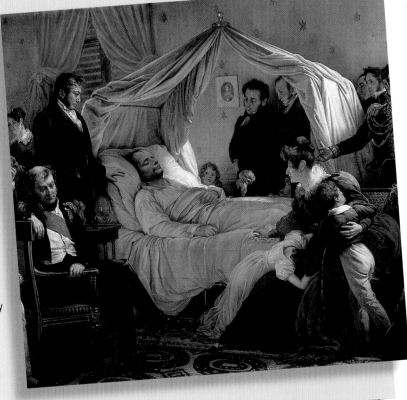

Acids and Bases

What do aspirin, grapefruit, and car batteries have in common? They all contain **acids.** An acid, which means "sour" in Latin, is a substance that releases hydrogen **ions** when it is dissolved in water. Some acids are weak enough for us to eat. Others are so strong they can burn through metal.

Bases are substances that can cancel out or neutralize acids. They produce hydroxide ions when they dissolve in water. Deodorant, lye, soap, antacid, and toothpaste all contain bases. Like acids, they can be weak or strong. Scientists measure the strength of acids and bases on a scale called the pH scale *(below)*. The "pH" stands for "power of Hydrogen." The scale runs from 1 to 14 and is based on the number of hydrogen ions the solution contains. A pH of 1 represents a strong acid and a pH of 14 indicates a strong base. Neutral substances such as distilled water register a 7 on the pH scale.

Plants and pH

Hydrangeas bloom with white, pink, or blue flowers. The color depends on the pH of the soil. In basic soil, pink flowers bloom. The same plant will have blue flowers in acidic soil and white ones in neutral soil.

The pH Scale

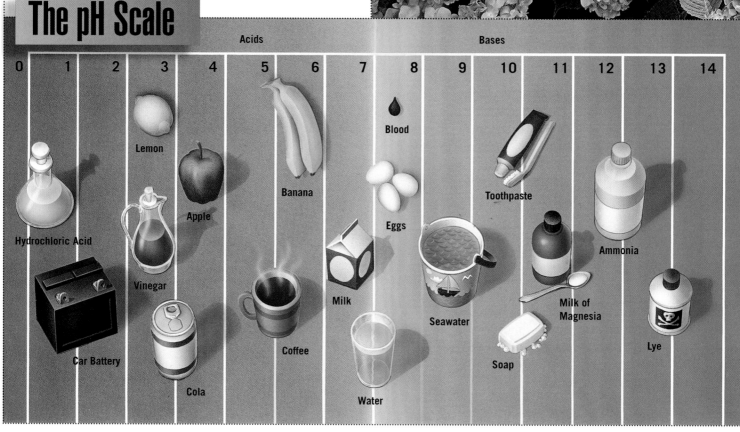

Acids   Bases

| 0 | 1 | 2 | 3 | 4 | 5 | 6 | 7 | 8 | 9 | 10 | 11 | 12 | 13 | 14 |

Lemon
Apple
Banana
Blood
Eggs
Toothpaste
Ammonia
Hydrochloric Acid
Vinegar
Milk
Seawater
Milk of Magnesia
Car Battery
Coffee
Soap
Lye
Cola
Water

Strange But TRUE!

Ant Acid

Stinging ants naturally produce a **chemical** called methanoic acid, a substance that is more commonly known as formic acid. This acid is a colorless liquid that gives off an irritating odor. A long time ago, people had to boil ants in a big pot in order to make formic acid. Today, it can be made from other chemicals in a laboratory. The acid is used in dyeing and finishing textiles and paper. It also plays a part in making fumigants, insecticides, and refrigerants.

Pickle Preservative

Before the invention of the refrigerator, people had to find other ways to preserve food. One popular method was pickling. Foods such as onions, peppers, and cucumbers were preserved in a jar containing vinegar, or acetic acid. The acid killed any bacteria and kept the food from going bad.

Measuring pH

How do you measure pH? One way is with indicators. Dyes from certain fruits and vegetables will change color when mixed with an acid or a base. The juice of a red cabbage *(right)*, for example, will be red in the presence of a strong acid, then will change to pink, purple, blue, and green as the solution becomes more basic.

The red cabbage indicator isn't very precise, though. Things with a pH greater than 7 or 8 will simply appear green. For a more exact reading, scientists sometimes use a universal indicator *(far right)*. This is a mixture of different indicators that changes color gradually over the pH scale, making it possible to read specific pH levels. For the most precise reading, however, scientists use a device called a pH meter.

Red Cabbage

Universal Indicator

14
13
12
11
10
9
8
7
6
5
4
3
2
1
0

Bitter and Sour Bites

Did you know that the stings from plants and animals are either basic or acidic? That means the sting can be neutralized to stop the pain. Add acid, such as lemon juice or vinegar, to a wasp's basic sting. Apply baking soda, a base, to a bee's or an ant's acidic sting.

How Strong?

Stomach Acid

To digest your food, the stomach releases gastric juices made largely of hydrochloric acid. At full power this acid can eat a hole in an aluminum tube *(right)*. The stomach acid is diluted but is still strong enough to burn a hole in a carpet. Luckily, your stomach has a protective coating of mucus. Even so, the acid causes the cells lining the stomach to wear out fast. The whole stomach lining is replaced about every three days!

CHEMISTRY 39

Organic Chemistry

Life could not exist without the element carbon. All living things contain it, even you. Because carbon is found in all living things, scientists used to think that only living organisms could produce **compounds** containing carbon. They thought it involved a mysterious "vital force." But by the early 1800s, scientists found out how to make carbon compounds in the lab. Today, of the thousands of carbon compounds we use every day, many are synthetic, or man-made.

Organic **chemistry** is the study of compounds containing carbon. The word "organic" means "coming from life." Organic chemists have a lot to study; 90 percent of all the known compounds in the world contain carbon. But not every carbon compound is considered organic. Calcium carbonate, carbon dioxide, and carbon monoxide, for example, are "inorganic" compounds. The study of these compounds and all other elements and their compounds is another branch of chemistry called inorganic chemistry.

Carbon is constantly being recycled in a process known as the carbon cycle. The illustration below shows the many ways this element, so crucial to life, is continuously absorbed and released. Plants absorb carbon from the air in the form of carbon dioxide during photosynthesis. They release this carbon dioxide when they are burned and when they die and decay in the soil. People and animals absorb carbon when they eat plants and other animals. When they exhale, expel waste, and decay after death, they release this carbon back into the environment as carbon dioxide and other forms of carbon.

Rocks absorb carbon and, after millions of years, form underground deposits of fossil fuels such as coal, oil, and natural gas. When these fossil fuels are burned in automobiles and factories, carbon dioxide is released into the atmosphere. That's how carbon that was once buried and stored underground is reintroduced into the cycle.

Where's the Carbon in Food?

Everything we eat contains carbon, the same element that's in pencil lead and diamonds. Take sugar, for example. Sugar is actually made up of three elements—carbon, oxygen, and hydrogen. An amazing thing happens when you pour sulfuric acid on sugar (don't try this!). The acid removes the oxygen and hydrogen from the sugar, leaving just the carbon (below, right). Now you can see exactly how much carbon there is in a little bit of sugar!

What Are Polymers?

Do you know you've got a bunch of polymers growing out of your head? The fibers in your hair are made out of lots of long chains of flexible **molecules.** These long chains are called polymers, and they all contain carbon. Other familiar polymers include rubber, silk, wool, plastic, and chewing gum.

Each year, the ocean takes a huge amount of carbon out of the atmosphere. It's absorbed by tiny ocean plants known as phytoplankton. But the ocean also returns a lot of carbon. Carbon-rich skeletons of dead marine animals settle on the ocean bottom and over time are compressed into limestone and oil.

Carbon Everywhere

The manufacture of organic compounds is one of the largest industries in the world today. It's easy to see why.

If you look around at the things you use every day, it's hard to find something that doesn't contain carbon. The fuel that runs the cars on the highway is an organic compound. The lettuce and vegetables in your salad at dinner are organic compounds. Dish detergent is an organic compound. Your clothes are made from organic compounds. Even the wood or plastic chair you're sitting on is made of organic compounds.

Below are some familiar objects that are made out of organic compounds. It's easy to see that carbon is everywhere.

Leather

Plastic

Food

Paper

What Are Forces?

Forces control how particles of matter interact. Physicists can use several words to define these **forces.** One is "observable"—you can see them at work in nature. Scientists also say that forces are consistent and measurable. A force may change, but it changes in an understandable way. Forces also have **magnitude** and direction.

There are two types of forces. The first are forces of physical contact—basically pushing and pulling—that can only come into play when matter touches other matter. There are also forces called "field" forces, often called "action at a distance," such as **gravity,** which work without any physical contact. The Moon, for instance, responds to Earth's gravity even though the two bodies don't touch.

Most of the time, more than one force is involved in every action. Forces that act in the same direction increase the effect of each force, as when the soccer player in the front at right kicks the moving ball, making it move faster. Forces acting in opposite directions decrease or cancel each other's effects, as when a player on the other team kicks the ball back.

Forces in Action

Forces are at work anytime something is put in motion. For instance, Mia Hamm of the U.S. Women's National Soccer Team demonstrates both forces and laws of motion (*pages 44-45*) as she beats the Chinese team to the ball in the photo above. Mia, the ball, and the other players all have mass, held on the Earth by gravity. **Friction** aids Mia when her feet push against the turf in the **acceleration** of those terrific sprints. The ball she follows is prone to inertia (*page 44*)— it will drop to the ground and lie motionless if Mia doesn't slam it away to score another goal.

Mass and Weight

Mass describes the amount of material in a body. Weight is the amount of gravitational force acting on the body. The astronaut on the Moon (*near right*) weighs less than he does on the Earth (*far right*) because the Moon's gravity is less. However, his mass is the same in both places.

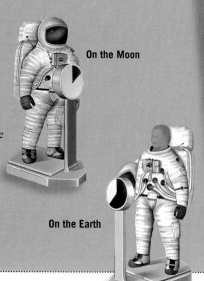

On the Moon

On the Earth

Friction

Some forces put objects into motion, but one—friction—slows them down. This is because every surface is a little rough. Rough surfaces catch on each other and slow a moving object. This can be a good thing; without friction, bicycle tires would just turn and slide, and a rider would never make it up a hill.

Gravity

Gravity controls the way separate masses are attracted to each other. The attraction is stronger when masses are large or close together (*far right*). It is weaker between smaller or more distant masses (*above, right*).

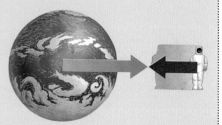

The Earth *(below)* has more mass than the Moon *(left)*, and so it pulls bodies to it more strongly. We say the Earth has a stronger gravitational field.

Compared with the Earth *(right)*, the Moon *(above)* has a small mass that attracts a person—or body—weakly. That's why astronauts can take big, bounding steps on the Moon.

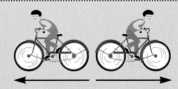

Let's Compare

Speed, Velocity, and Acceleration

Speed is simply a rate of travel.

The same speed, but two different velocities!

In physics, **speed, velocity,** and acceleration are all words used to describe different aspects of motion. Speed means a rate of travel over distance. Velocity is a measure of speed in a certain direction. And acceleration refers to a change in velocity: An object or a person goes faster or slower or in a different direction.

Acceleration means speeding up, slowing down, or changing direction.

Newton's Laws of Motion

Why do box cars speed along the way they do? Why do they stop? Can we predict how they will move? We can answer these questions today thanks to the English mathematician Isaac Newton (1642-1727). Before the 17th century, no one had analyzed motion. People believed that all objects just naturally remained in one place, unless pushed or pulled. In Italy, Galileo challenged this idea with experiments and questions, but Newton explained it in a formula.

Newton's three laws of motion, published in 1687, showed that the way things move can be described in simple mathematics. The first law states that moving objects continue to move in a straight line, and objects at rest remain at rest, unless moved by an outside **force.** The second law describes how a force acting on an object causes **acceleration.** The third law states that for every action there is an equal and opposite reaction. Using his laws of motion, people can predict the movement of the planets—and design faster box cars.

What's Inertia?

Objects do not move on their own unless a force sets them in motion. Once they are in motion they remain in motion unless a force stops them. This resistance to change is called inertia. When a bicycle suddenly stops at an obstacle, as in the photo at left, the rider will continue to move forward and tumble over the top. The same is true when a bus stops: People continue to move forward, unless they hold on to something to steady themselves.

Mooove!

This race shows all of Newton's laws in action. 1. The box cars stand at rest unless something (a human body) sets them in motion. Once moving, they keep moving until **friction** stops them. 2. The speed of the cars' movement forward depends on how their **mass** is accelerated. 3. The force pushing the car forward is equal to the force applied backward—the feet pushing against the ground.

First Law of Motion 1

This law states that objects at rest tend to remain at rest, unless set in motion by an outside force. Also, objects in motion tend to stay in motion in a straight line, unless affected by an outside force. This cart is a good example: It will stand still unless the horses pull it. It will keep rolling once it starts, until friction or the horses halt it.

2 Second Law of Motion

This law shows how force, mass, and acceleration are related: force = mass x acceleration. Diagram 1 shows how a small force moves a small mass as far as point A. In 2, a greater force moves that mass to point B. Double the force and you can double the mass moved to point A, as in 3. And 4 shows you what happens if the mass is doubled and the force isn't.

Third Law of Motion 3

Newton's third law states that every action has an equal and opposite reaction. With each mighty stroke the rowers at right apply the same force backward on the water as the force that pushes the boat forward. In other words, any action in one direction will be balanced by a reaction in the opposite direction.

Newton's Law of Universal Gravitation

On their way to the Moon, Apollo astronauts publicly thanked Isaac Newton, who made their trip possible. Before Newton's ideas, the fact that objects fell to the ground was thought of as a property of the objects themselves—like their textures or colors. Newton's insight was that **gravity** is a universal **force.** It is measurable and predictable and felt not only by matter on Earth, but also by all the heavenly bodies.

Newton worked out a mathematical formula that could predict the pull of gravity on any object. His formula states that the gravitational pull between any two objects in the universe is directly proportional to the product of their **masses,** and that gravitational force is inversely proportional to the square of the distance between two masses. So a planet the size of Earth but twice as far from the Sun feels only a quarter as much gravitational pull from the Sun as Earth does. The same planet four times as distant will feel only $\frac{1}{16}$ as strong a pull. Newton's formula is still used today to design space shuttle missions.

Objects in Orbit

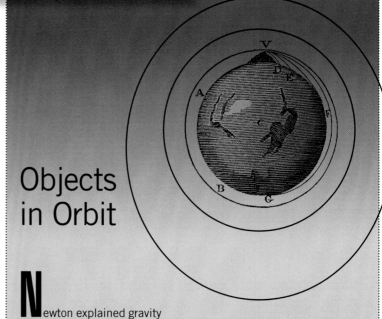

Newton explained gravity and **acceleration** with this diagram of a cannon being fired from an imaginary mountaintop above Earth's atmosphere, where air **resistance** does not slow the ball down. When the cannon fires, the ball travels quite far. If a stronger charge were used, shooting the ball out at a greater speed, it would travel farther. If a cannon could accelerate a ball enough, Newton said, it would travel so far that it would miss Earth's surface entirely, and go into orbit.

Falling Objects

Stories tell that when Newton saw an apple fall from a tree, he realized that the same **force** that brought the apple to the ground held the Moon in its orbit. But why doesn't the Moon fall to Earth? Newton theorized that it was because the Moon's distance from Earth reduced the force of attraction between them. The Moon is falling toward Earth, but the pull of gravity is not strong enough to counteract the Moon's forward motion—which is in a straight line—so that the Moon's path ends up pulled in as an **orbit** around Earth.

Outside Earth's Gravity

Since human bodies don't adjust well to weightlessness, spaceships and space stations of the future might need artificial gravity. In this design for an interstellar explorer, the acceleration of the spinning cylinder acts like gravity for the crew within. The fuel needs of a multiyear voyage are met by a nuclear reactor, attached at a safe distance from the living quarters. The ship would not use much fuel once it was away from a planet's gravity field.

Defying Gravity

At 265 km (165 mi.) above Earth, astronaut Bruce McCandless can float freely. His mass is too small to pull him or his spaceship down to Earth. The space shuttle *Challenger* orbits just a few feet away.

People

Galileo

In the 1500s, Italian scientist Galileo Galilei experimented with motion. An inclined plane and bronze balls *(below)* showed that a falling object accelerates at 9.8 m (32 ft.) per second for every second it falls. He also proved that objects in a vacuum fall at the same rate, no matter what their weight.

Would **You** *Believe?*

Antigravity Car?

It sounds like a science-fiction tale, but scientist Ning Li of the University of Alabama is building an antigravity device. Based on the physics of spinning disks in a very cold chamber, Li's machine can reduce the weight of small objects placed above it. She hopes to design an antigravity car by 2010. If she succeeds, she will have objects floating and no longer being pulled to Earth.

Try it!

Galileo's Experiment

Try one of Galileo's experiments to see if objects with different weights will fall at the same rate. Galileo dropped cannonballs from the Tower of Pisa. Stand on a chair and drop two similar objects onto a cookie sheet at the same moment: Did you hear them land at the same time?

What Is Energy?

Kinetic Energy

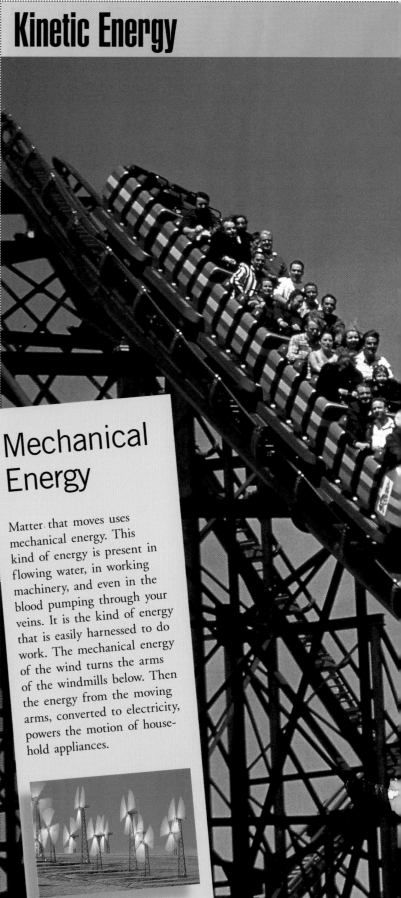

When you say "I'm full of **energy** today! Let's go out and do something!" you're close to the concept of energy as it is defined in physics. Energy can be described as the ability to do **work**—that is, work in the physics sense *(page 51)*.

Energy is continually being stored and converted into many forms throughout the universe. A roller coaster, for example, gains gravitational potential energy as it climbs, and converts it into kinetic energy as it drops. The total amount of energy and **mass** in the universe never changes. When it seems the energy in a system has been used up, as when a flashlight battery dies, that energy hasn't disappeared; it has been transformed into light shining from the flashlight. Science calls this principle the law of conservation of energy.

Almost all energy on Earth arrives in the form of heat and light from our Sun. The five forms of energy in the universe are chemical, mechanical, heat, electromagnetic, and **nuclear.**

Chemical Energy

Chemical energy is the energy involved in chemical reactions. Eating an ice-cream bar gives you chemical energy as your digestion converts the sugar to fuel for your body. Every time you take a breath, you're taking advantage of the chemical reactions between oxygen and elements in your bloodstream to process body wastes. Cars burning gasoline are another example. Their engines change the energy of a chemical **reaction** into motion in the engine and wheels (and lots of heat and noise).

Mechanical Energy

Matter that moves uses mechanical energy. This kind of energy is present in flowing water, in working machinery, and even in the blood pumping through your veins. It is the kind of energy that is easily harnessed to do work. The mechanical energy of the wind turns the arms of the windmills below. Then the energy from the moving arms, converted to electricity, powers the motion of household appliances.

Scientists think of mechanical energy as divided into two types. Potential energy is energy stored in a system when work is done. Kinetic energy is the energy of objects in motion. At the top of the rise, these roller coaster cars are packed with potential energy that came from the work of raising them to that height. When the cars swoop down, potential energy is changed to the kinetic energy of a hair-raising ride.

Electromagnetic Energy

The light all around you is electromagnetic energy. So is the electricity that comes through the power lines. Radio and x-ray waves are forms of electromagnetic energy. Its most impressive form is the lightning flash!

Nuclear Energy

When atomic nuclei split or join together, heat and electromagnetic energy are released. This is the source of **nuclear** energy. Nuclear reactions produce the heat and light of the Sun and all the other stars.

Try it!

Heat Energy

Transform heat energy into mechanical energy. Put an empty bottle no bigger than 354 ml (12 oz.) in the freezer for 15 minutes. Take it out, dip the mouth in water, and set its cap on top upside down. Wrap both hands around the bottle and hold it upright. In a few minutes the cap will bob up and down. Heat energy from your hands made air expand and push out of the bottle, using mechanical energy to move the cap.

Simple Machines

Early humans discovered the principles of machines when they first put the **forces** of the natural world to use. They pried up a heavy rock with a long branch and with that invented the lever, or they used a wedge-shaped stone to cut meat.

A machine doesn't have to be a complicated collection of gears and wires. Any device that sends a force or changes it is a machine. The first machines are often called "the six simple machines," though some of the six are just special uses of others. The basic six are the inclined plane; the screw—really just an inclined plane, wrapped around a cylinder; the wedge; the lever; the wheel and axle; and the pulley. An inclined plane—a ramp—is at work in the photo below. It spreads out the effort needed to lift the box up into the truck. The same work clearly takes more effort in the photo above, right.

You have to use a certain amount of force to lift a load. If you pull it straight up or lift it up, as in the picture at right and the diagram below, you will need a lot of force over a short distance.

The Inclined Plane

If you lift a load over an inclined plane, as the two people are doing at left, you will use the same amount of force as in the photo above, but you apply it over a longer distance. As a result, you will need less of an effort at each point along that distance.

Levers

Levers are bars that move on a fixed point, or pivot, called a fulcrum. The fulcrum balances two forces: an effort force and a **resistance** force, or load. The position of the fulcrum changes the way the lever uses force applied to it. The closer the fulcrum is to a point of resistance, the more force it can apply against that resistance. If the fulcrum is far away from the point of resistance, it can apply less of the force against resistance. When the ancient Greek mathematician Archimedes wanted to show the effort-saving use of the lever, he said, "Give me a place to stand on, and I will move the world."

1st Class

Force   **Fulcrum**   **Load**

In these levers, the fulcrum lies between the applied force and the point of resistance. The scissors' screw is the fulcrum.

2nd Class

The fulcrum in this type of lever is fixed at one end, with the force applied at the opposite end. The resistance, or load—in this case, the nut's shell—stands between them.

Force   **Load**   **Fulcrum**

3rd Class

Load   **Force**   **Fulcrum**

Here the force (the elbow) lies between the fulcrum (the shoulder) and the load (the tennis ball). This type of lever magnifies the distance while it minimizes the amount of effort required.

Wedge

A wedge is really two machines —two inclined planes placed back to back. Long, thin wedges split or raise objects with less effort than short, thick wedges, because their inclined planes are longer. Knives, plows, axes, and the wood-splitters shown here are wedges. Metal zippers use a tiny wedge in the slide to force the interlocking teeth apart when you pull the slide down. Pull it up, and two more wedges in the slide press the teeth together again. But try pulling a zipper apart by hand. Then you'll appreciate the power of the wedge!

What's Work?

You might think of work as chores, or as a paying job, but in physics the term has a special meaning. When a force moves an object in the direction of that force, that is **work,** scientifically speaking. The boy at top right might be surprised to find he's not doing work, since his arms use force to hold the books. But when he lifts the books off the floor *(right)*, he is doing work: The books move in the direction of the force he applies.

Complex Machines

Gears

Most modern machines, no matter how complicated they look, can be broken down into simpler machines working together. Many of the most useful machines involve wheels.

A key moment in the history of machines arrived when someone realized that you could attach a small wheel to a larger one with an axle. When the large wheel turns, the small wheel will turn with great **force.** If you turn the small wheel, the large wheel moves around a long way, with very little force needed. This is the principle behind the use of wheels as gears.

Wheels and cylinders also make things easier by reducing **friction,** as the ancients discovered when they pulled heavy blocks over rolling logs. All machines, from simple to complex, have the same goal: to make things happen with the least effort possible.

Even the largest system of gears is simply circles of different sizes joined together so that movement of one affects the movement of another. In freewheel bicycle gears, a chain connects a large circle in the hub at the pedals to smaller circles at the rear wheel. Pushing the pedals turns the large circle, which spins the small circle several times, moving the rear wheel of the bike with it. Changing gears changes the size of the rear gear wheel, and thus the distance it turns.

Many Machines in One

Since the 19th century, bicycle makers have added more and more machines to their bicycles to make them easier to ride. The gear systems on modern bikes add the most oomph, but see what else is used here.

Lever

Lever

Wheel and Axle

Gear

Gear

Wheel and Axle

Pulley System

Wheel and Axle

They're everywhere! A large circle (wheel) connected to a smaller circle (an axle, a wheel disguised as a cylinder) links distance traveled and power needed in an incredibly useful way. Screwdrivers, doorknobs, faucets—and the antique car chassis at right—are all wheel-and-axle systems.

Pulleys

A pulley is a wheel with a rope moving around it. A fixed pulley *(right)* changes the direction of force, making work easier because it is easier to pull down than push up. Movable pulleys move with the load. Each section of string supporting the movable pulley system decreases the

force needed to move the load. The pulley at left, with two sections of string, cuts the force needed in half. If the first pulley hangs from another, making four sections of string, only a quarter as much force will be needed to lift the load.

Try it!

Build a Crazy Machine

Students from the University of Texas won first prize in the 1997 Rube Goldberg contest to build a complicated machine for a simple task. Their machine could load a compact disk into a CD player in 35 steps. The winning entry included a windmill, a waterwheel, a mousetrap, and a pendulum.

Rube Goldberg: No More Oversleeping

Inventors have long tried to design machines to do more than one job. Cartoonist Rube Goldberg, on the other hand, excelled by drawing machines that made a simple task more complicated. His cartoon at right shows a series of levers and pulleys being used to wake a person in nine steps.

When sun comes up, magnifying glass (A) burns hole in paper bag (B), dropping water into ladle (C) and lifting gate (D), which allows heavy ball (E) to roll down chute (F)—Rope (G) lifts bed (H) into vertical position and drops you into your shoes (I).

P.S. You can't go back and sneak a few winks because there's no place to lie down!

What Is Electricity?

Try to imagine what your world would be like without electricity: no lights at the flip of a switch, no TV, refrigerator, or computer. But electricity is not just a modern convenience; it is also a powerful force of nature, as when a bolt of lightning streaks across a stormy sky. To understand how electricity works, we need to look at **atoms.**

Atoms consist of three basic particles: positively (+) charged **protons,** negatively (-) charged **electrons,** and **neutrons,** which have no charge *(below).* The protons and neutrons are grouped together in the **nucleus,** or center, of an atom. The electrons spin around the nucleus at high speed. Normally, each atom has an equal number of protons and electrons. The positive and negative charges cancel each other out, leaving the atom "electrically neutral," or uncharged. But when an atom gains or loses an electron, we start to see electricity at work!

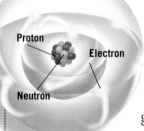

Proton
Electron
Neutron

What's in a Name?

Electron

When fossilized tree sap called amber is rubbed with a cloth, it can attract and pick up light objects such as feathers. The ancient Greeks discovered amber's magical ability, but they did not understand what caused it. A 16th-century English physician named

William Gilbert found several other materials that reacted just like amber when rubbed. He called them "electrics," from the Greek word *elektron,* meaning amber.

IT'S THE LAW

The Law of Electric Charge

There are only two kinds of **electric charges:** positive (+) and negative (-). Charge affects the way objects behave when they come near each other. They will exert one of two different electric **forces:** attraction (pulling toward) or repulsion (pushing away). If the charges are opposite, they will attract each other *(top right),* if alike, they repel *(bottom right).* This idea that opposites attract and identical charges repel is so important to understanding electricity that it became known as the law of electric charge.

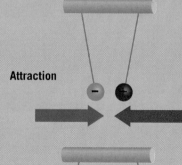

Attraction

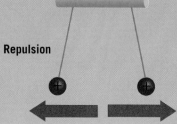

Repulsion

The Power of Electricity

Electricity not only lights up this busy Hong Kong street, it also holds together the signs, buildings, people, and everything else in the entire universe! All matter is made up of atoms, which are bound together by the electric force of the attraction between protons and electrons. Without this force, nothing could exist.

Body Electric

Snowboarding, doing homework, even keeping your heart beating, all require electricity. Your body's cells use electricity to communicate with each other. No specific organ makes electricity; it is produced chemically by every cell. Nerve and heart cells, in particular, make a lot of electricity. Although individual cells generate only a tiny amount, if you added together all your body's trillions of cells, the total electric output would light a 40-watt bulb.

Electric Eel

Electric eels can generate, store, and discharge electric current. These eels from South America's Amazon River can reach 2.4 m (8 ft.) in length. Their tails are packed with special cells that generate electricity. They use this current at low **voltage** to navigate the muddy water, locate food, and communicate. But they can discharge up to 600 volts for defense or to stun and kill prey.

Try it!

Feel the Force

Around any charged object, there is an invisible area called an electric field. Within this field, the object's electric charge exerts a force. You can feel the force of electricity yourself with this simple experiment: Blow up two balloons and rub them with a wool cloth. Then get a pair of "clingy" socks fresh out of the dryer. The balloons represent electrons and the socks are protons. The socks will stick to the balloon, just as protons are attracted to electrons. Next bring the two balloons slowly toward each other and watch how they repel each other.

Attraction

Repulsion

Static Electricity

Have you ever walked across a carpet, then received a shock from touching a metal doorknob? If so, you have experienced **static electricity.** The term static, which means unmoving, describes a special form of electricity that occurs when **electrons** are transferred from one object to another without any further movement. This is in contrast to the continuous flow of electrons that creates an electric **current** *(pages 60-61).*

The electrons within the **atoms** of certain materials are not held tightly in their orbit; they easily break free when the material is rubbed against another object. When these "free" electrons leave an atom, the atom takes on a positive **charge** because it now has more **protons** than negatively charged electrons. The atom that the electron joins then becomes negatively charged, because it contains more electrons than protons.

When a comb or balloon is run through hair, it causes electrons to jump from the hair onto that object, leaving each strand positively charged. Since like charges repel, the strands of hair try to get as far away from each other as possible, with "frightful" results.

Magic Tricks with Electricity

For centuries, static electricity was a mystery to scientists, who could see its effects but had no idea what caused it. This sparked many imaginative demonstrations, such as the human chain *(above),* devised by 18th-century English

experimenter Dr. William Watson. The static **electric charge** from the woman's hand was transferred to the boy's feet, then through his body to the young girl's hand, which attracted lightweight particles of chaff from the table.

Law of Conservation of Charge

IT'S THE LAW

Electric charge is neither created nor destroyed, it is only moved around. For example, when you rub a glass rod with a piece of silk, free electrons jump from the glass to the cloth. The rod loses a certain number of electrons and takes on a positive charge. The silk, which gained exactly the same number of electrons, becomes negatively charged. No protons or electrons were created or destroyed by the exchange. This is known as the law of conservation of charge.

How Objects Get Charged

Rubbing two objects together is one way to charge them. This is known as charging by **friction**. It is not the movement or friction that causes electrons to jump. Rather, rubbing merely allows more points of contact between two objects than simple touching.

Another way to charge something is called induction, and it does not involve contact at all. When a charged object is brought near a neutral object, its electric charge can force electrons on the neutral object to rearrange their positions, making it either positive or negative.

1

When you rub a balloon with a piece of cloth, the cloth loses electrons and the balloon gains these electrons. The balloon now has a negative charge because it has more electrons than protons.

2

As the negatively charged balloon comes near a wall, the extra electrons on the balloon repel electrons in the wall— since like charges repel— leaving that area of the wall's surface positively charged.

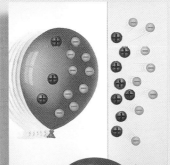

3

The positively charged wall now attracts the negatively charged balloon. The balloon will stick to the wall until the charges—or the number of protons and electrons—in both objects even out.

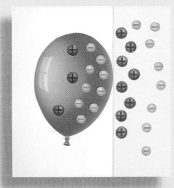

Determining Charge

Atoms in some materials hold their electrons tightly in orbit, and atoms in others hold their electrons more loosely. Scientists have ranked materials in order of their tendency to keep or give up their electrons. This ranking is called the triboelectric series, from the Greek word *tribein,* which means to rub. A brief list is shown at right.

When two materials are rubbed together, the one higher on the list will give up its free electrons, becoming positively charged, while the lower one will become negatively charged. The farther apart the two materials are on the list, the greater the static charge that will be built up between them.

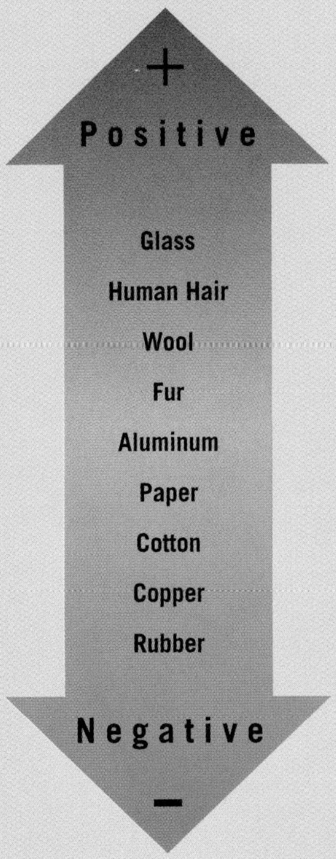

+
Positive

Glass

Human Hair

Wool

Fur

Aluminum

Paper

Cotton

Copper

Rubber

Negative
—

Would **You** *Believe?*

Putting Static Electricity to Work

Have you ever wondered how a photocopying machine works? The secret of getting the ink to stick to the paper is static electricity. First, the original page is illuminated with a bright light, then the image of the page is projected onto a negatively charged metal drum. The electric charge remains only where the dark parts of the image strike the drum. The fine black toner powder, which is positively charged, is attracted to the

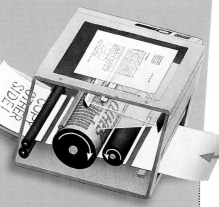

Photocopier

dark areas of the drum. The copy paper is then rolled over the drum to pick up the image and heat-seal the toner permanently to the paper.

Static Discharge

When an object becomes charged by **static electricity,** its normal balance of **protons** and **electrons** is upset. The only way to restore the balance is for some of the object's electrons to jump to another object of opposite **charge.** This loss of static electricity, or **electric charge,** is called static discharge.

Static discharge often occurs without being noticed; the extra electrons move slowly onto other objects or escape into water **molecules** circulating in the air. But static discharge can also be rapid and accompanied by a shock, a crackling noise, or even a spark.

Around the house, this kind of sudden discharge is more common in winter, when the inside air is drier and electrons cannot scatter safely into surrounding water vapor. The ultimate and most spectacular form of static discharge is lightning *(right).*

People — Benjamin Franklin

In 1752, Benjamin Franklin, assisted by his son, flew a kite into a thundercloud to prove his theory that lightning is a form of static electricity. As soon as the kite entered the cloud, the fibers on the kite string stood erect, just as they had when charged with static electricity in Franklin's laboratory. Suddenly, a spark jumped to his finger from a metal key tied to the end of the string. Franklin had made an important discovery and was fortunate enough to live to tell the tale.

Ultimate Discharge

Friction between water molecules builds up strong negative charges in the base of a storm cloud. The ground below the cloud then becomes positively charged.

The negative charge attracts the positive charge, creating a massive spark that is seen as a flash of lightning.

A Shocking Experience

Walking across a carpet can give your body an electric charge. As your shoes rub against the fibers, they pick up electrons. These electrons move through your body and you become negatively charged. This negative charge goes unnoticed until you touch something, such as a metal doorknob. The extra electrons in your body are attracted by the protons in the doorknob. They jump from your hand to the knob's metal surface with a sudden spark like a miniature bolt of lightning.

Ouch!

Lightning rods are placed on the roofs of houses and the tops of other tall structures as a safety precaution. They work on a principle called grounding, which relies on the fact that the earth, or ground, is so large that there is plenty of space for an electric charge to spread out and disappear. When lightning strikes the rod, a wire connected to it safely carries the electricity into the ground. Some skyscrapers, such as the Empire State Building, are hit by lightning hundreds of times every year. Benjamin Franklin tested the first lightning rod shortly after his famous kite-flying experiment. His invention later saved his own home from destruction.

Lightning Rod

Ground

Then & NOW!

Van de Graaff Generator

This huge machine was built in 1931 by American physicist Robert J. Van de Graaff and is still in use today. It generates 2.5 million volts of artificial lightning that passes around the metal-framed cage while the operator sits safely inside.

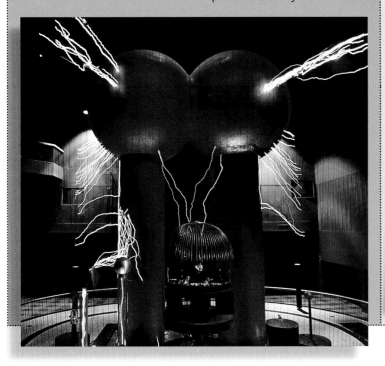

Imagine That!

Spark of Life

In 1816, a teenager named Mary Shelley wrote a novel about a doctor who creates a monster out of dead-body parts. She described a "powerful engine," probably an early battery, that gave the creature life. The 1931 movie version of Shelley's *Frankenstein (below)* used sparks from a giant static electricity generator.

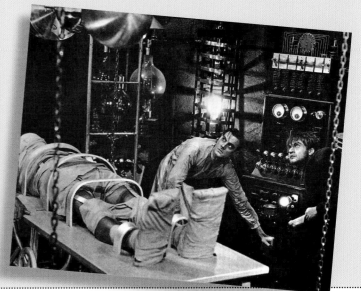

Current Electricity

E lectricity was once thought to be a kind of invisible liquid. Scientists now know differently, but we still refer to the flow of **electrons** through a wire as electric **current.**

Force, supplied by a power source such as a battery, is needed to make the electrons move. This force, or "electric push," is called **voltage** and is measured in units called volts (V). It is named for Alessandro Volta. The rate at which the electric current flows is measured in amperes (A) and is named in honor of André-Marie Ampère. Current flows because there is a difference in voltage between one end of the wire and the other. This is similar to how water flows from a high to a low point. When voltage is applied to the end of a wire, each free electron is pushed just a little by this force, and the charge moves down the line. Although each electron moves only a little bit, the impact of this continuous movement allows the energy of the electric current to flow through the wire at more than half the speed of light—practically instantly.

How Current Flows

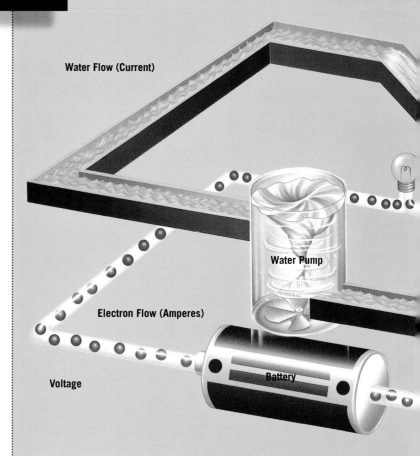

Water Flow (Current)

Water Pump

Electron Flow (Amperes)

Voltage

Battery

People   Luigi Galvani

L uigi Galvani discovered electric current by accident in 1786. The Italian professor of anatomy was dissecting a frog with a steel scalpel when his assistant happened to produce a spark with a static electricity machine. Suddenly, the frog's leg twitched.

Fascinated, Galvani conducted many more experiments and concluded that the twitching was caused by what he called "animal electricity."

Let's Compare

Conductor vs. Insulator

Insulator

Conductor

E lectric current moves easily through certain metals, such as copper and silver. They are known as **conductors.** Other materials, like rubber and glass, are called insulators because they don't conduct electricity very well. Conductors and insulators are often used together. For example, copper wire is covered with a layer of rubber so that the electricity will stay in the wire and flow only where it is supposed to go.

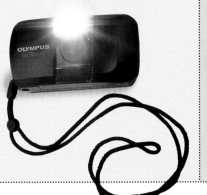

Water Slide

Light

A good way to understand the flow of electric current is to compare it to the flow of water. A battery, like a water pump, supplies the force—voltage, in the case of electricity—to move the current. Whereas the flow of water is measured in units like liters or gallons per minute, the flow rate of electrons is calculated in amperes, or amps for short.

The current moves along at its full work potential until it encounters a change, or need for some of that energy. Just like the water slide in the illustration at left, an electric light uses up some of this energy and reduces the power of the current. It must return to the battery, or water pump, to regain its peak work potential.

Then & NOW!

Capacitor

A capacitor is a device that stores an **electric charge** and discharges it all at once. The Leyden jar *(above, right),* invented in the 1740s, was an early example. Capacitors are still used today. A good example is a camera flash. A capacitor stores an electric charge from the camera's batteries and when needed, sends a big surge of current instantly to the flashbulb.

Volta's Invention

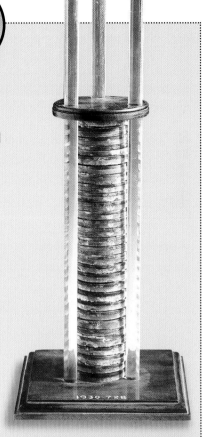

In 1800, Alessandro Volta, an Italian physics professor, made an important discovery. By placing a pad soaked in a weak acid solution between disks of two unlike metals, he found he could generate a steady supply of electric current. Volta called this unit a cell, and he soon learned that the more cells he piled on top of one another, the more power was generated. He had invented the world's first practical battery, called a voltaic pile.

Batteries

Batteries provide us with a cheap, reliable—and portable—source of electric power. Tiny button cells run watches and hearing aids. Larger AAs, Cs, and Ds can be found in almost every home. Milk-jug-sized batteries provide the power to start our cars.

Like Volta's earliest example, many batteries are still made of smaller units called cells.

The voltage output of a typical one-cell D battery *(above)* is 1.5 volts. But if you stack two of them in a flashlight, the combined cells produce 3 volts. Sometimes several cells are all stacked inside one battery, like the 9-volt one at left. This battery has six separate cells, each having 1.5 volts of power.

Resistance to Electricity

Imagine a marble dropping through the air. It falls easily, with little **resistance.** But if the marble fell into a container of honey, its momentum would be slowed considerably by the honey's thickness. Like the marble in the honey jar, there is an opposition to the movement of **electrons** in a conducting wire that keeps them from flowing freely. This force is called resistance, and any material that opposes electron flow is called a resistor. Resistance is measured in units called ohms.

Resistance is the thief of electric **current;** every day in the United States, a tenth of the electricity transmitted through power lines is eaten up by resistance, because the electron flow uses up energy as it passes through the wire. But resistance also provides many benefits. Appliances we use regularly, such as hair dryers and electric stoves, rely on resistance to generate heat. And the light bulb couldn't light up without resistance.

IT'S THE LAW

Ohm's Law

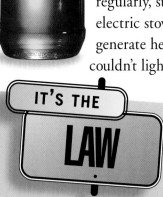

While conducting experiments with electricity, a German schoolteacher named Georg S. Ohm found that something was reducing the **voltage** flowing through a wire. He theorized that even a good **conductor** of electricity, like copper wire, resists the flow of **electric charge.** He came up with a mathematical formula to explain the relationship of the various forces at work: voltage = current x resistance. In his honor, it was named Ohm's law.

What Causes Resistance?

To understand resistance and how it works, imagine that your school hallway is a wire and you and your fellow students are electrons flowing through that wire. If the hall is long and narrow, it will take you longer to get from one class to another than it would take in a wide, short corridor. Furniture, lockers, or other obstructions in the hallway will also slow your progress. And if the inside temperature is too high, the students become hot and sticky and don't behave or move as orderly as they normally would.

Like the school hallway filled with students, the flow of electrons through a wire depends on four main elements that cause resistance: the length, diameter, temperature, and composition of the conductor. The longer the wire, the more obstacles the electrons can encounter. A thin wire offers less space for the electrons to travel through at a time. At higher temperatures, atoms, like people, move more randomly and get in the way of the flow of electrons. A good conductor such as copper or gold is like an empty hallway that offers little resistance to the movement of free electrons through it.

Bright Idea

Thomas Alva Edison, an American inventor, lit up the world with the first practical electric light bulb. He and his assistants spent well over a year testing thousands of different materials to use as a filament inside the vacuum glass bulb. The filament had to provide enough resistance to glow brightly when a current passed through it but not burn up. A short piece of carbonized, or scorched, sewing thread proved able to withstand the intense heat. On October 22, 1879, his electric light bulb was lit and lasted 13½ hours.

Thomas Edison displays his invention.

A Human Conductor

Bill Wysock may seem to be tempting fate by sitting on a 1.5 million-volt electric transformer. But at very high frequencies *(page 75)*, electric current flows only on the surface of a conducting material, in this case the surface of his body. The current then passes harmlessly through his hands, not only lighting a 40-watt bulb but also traveling up the rod to generate mini-bolts of lightning.

Overcoming Resistance

In 1911, Dutch physicist Heike Kamerlingh Onnes proved that certain metals allowed electric current to flow without resistance if their temperature was lowered as close to absolute zero, 0° Kelvin *(page 24)*, as possible. These so-called superconductors offered a way for electric systems to work more efficiently and cheaply. It has also been found that superconductors can repel magnets, like the disk levitating the cube magnet below. Today, scientists are developing new ceramic materials that will even superconduct at higher temperatures.

Temperature affects the flow of electrons moving down a wire. In a cool wire, the electrons vibrate less, so the current flows more energetically *(above)*

Electrons can move more easily through a short, thick wire than through a long, thin wire.

Electric Circuits

Every time you flip a light switch, you assume that the light will come on. But for the system to work, there must be a complete path, or loop, that lets the **current** flow from the power source to the bulb and back. This continuous path is called a circuit. There are two main types of circuits: a simple series circuit and the more complex parallel circuit.

When all of the various parts of the circuit are connected to each other, without gaps, as shown in the illustration below, it is called a closed circuit, and the electric current can flow.

If the path is broken anywhere in the circuit—for example, if the light bulb burns out—the current flow will be interrupted. This is called an open circuit.

Parts of a Circuit

A simple electric circuit has four main parts: a power source, like the battery shown below; a path, usually a conducting wire, to carry the current through the circuit; a load, such as a light bulb, that requires electricity; and a control, or switch, to turn the current on or off. When the switch is off, the circuit is open and the current cannot flow. Turning the switch on closes the circuit, allowing the current to flow and light the bulb.

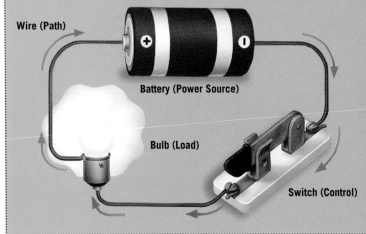

Wire (Path)

Battery (Power Source)

Bulb (Load)

Switch (Control)

In a sense, these cars on a highway are like **electrons** flowing down a wire, and the tollgate is like an electric switch. The cars continue to move down the road as long as the gate is open. Closing the gate is like turning off the switch; it interrupts the circuit and stops the flow of current.

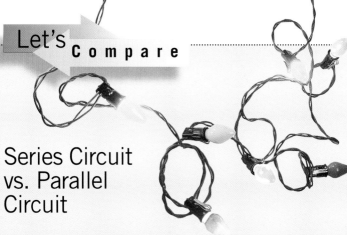

Series Circuit vs. Parallel Circuit

In a series circuit *(right)*, there is only one path the current can take to flow through the loop. If the path is broken—for example, if one bulb burns out—the current cannot flow and none of the other bulbs will light.

In a parallel circuit *(bottom right)*, the wire leaves the battery and branches off into a separate path to each bulb. If a bulb burns out, the circuit remains complete to the other bulbs and the current still flows. Strings of Christmas tree lights used to be wired with series circuits; one bad light and the whole string went dark. Today they are made with parallel circuits, and burned-out bulbs are easy to spot.

Series Circuit

Parallel Circuit

High-Wire Act

Did you ever wonder why birds can sit on a high-voltage wire and not get shocked? When a bird perches on just one wire, it is not making a complete circuit. The current passes through the wire, and not the bird, since the **voltage** is the same all along the wire. However, if the bird's feet or wings touch two different wires, each with different voltages, the bird creates a path between the voltages. This path happens to include the unfortunate bird.

Then & NOW!

Controls in a Circuit

The devices we use to control the flow of current through a circuit have changed greatly in the last 100 years. Vacuum tubes—sealed glass containers with all the air removed—were invented in 1904. They were large and fragile and grew very hot while they were being used. The transistor was introduced in 1948. It was much smaller and faster, and it used a lot less power. Today, the tiny microchip contains the entire electric circuit, including millions of transistors.

Microchip

Transistor

Vacuum Tube

Electric Power

Many of our home appliances need so much **energy** that it would be impossible to run them with batteries. It would take about 1,000 flashlight batteries, for example, just to toast a piece of bread. Ovens, hair dryers, and water heaters, too, require more power than a battery can deliver. The amount of power any electric device uses is measured in a unit called a watt (W). It was named in honor of James Watt, a Scottish engineer who invented the steam engine in 1769.

Electric power is generated at power plants. The electricity is sent for long distances over transmission lines, or power lines, to our homes. For the most part, the electric outlets in North American houses are fitted for 120 volts. In many other parts of the world they are 220 volts.

What's a Watt?

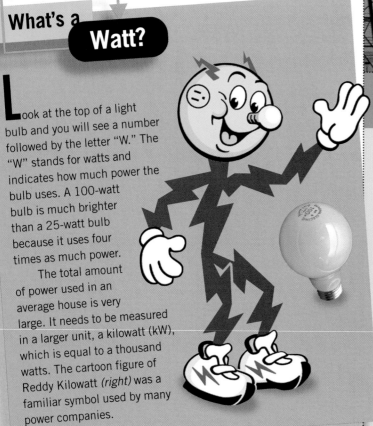

Look at the top of a light bulb and you will see a number followed by the letter "W." The "W" stands for watts and indicates how much power the bulb uses. A 100-watt bulb is much brighter than a 25-watt bulb because it uses four times as much power.

The total amount of power used in an average house is very large. It needs to be measured in a larger unit, a kilowatt (kW), which is equal to a thousand watts. The cartoon figure of Reddy Kilowatt (*right*) was a familiar symbol used by many power companies.

Household Circuit

Your house is like a giant parallel circuit. This is because if it were wired like a series circuit, every time you turned off one light, all the electricity would go off all over the house. Hidden behind the walls is a maze of wires, organized in parallel circuits that safely carry just the right amount of electric power to various parts of the house.

Once the power enters your house, it passes through an electric meter, which measures the kilowatts used per hour. The power then continues through a circuit breaker or a fuse box, a safety device that prevents electric hazards such as fires. From there, it branches off to supply the power needs of the house. A large appliance, such as an electric stove, uses so much electricity that it has its own circuit (*red*).

Carrying Power

High-**voltage** power lines are held safely overhead and out of reach by tall towers, called pylons. They can carry as much as 765,000 volts of power in each wire. It is less expensive to generate electricity at one central location—a power plant—and then to send it over long distances, even though a lot of the power is lost due to **resistance** in the wires.

What's a Short Circuit?

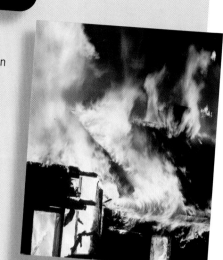

A short circuit happens when too much **current** flows across two wires. The wires can then become red hot and cause a fire. Fuses and circuit breakers are built into a home's main circuit to prevent this. If too much power suddenly surges through the circuit, the fuse blows or the breaker trips, opening the circuit and shutting off the current.

Let's Compare

AC & DC Currents

Direct current, or DC, flows in only one direction. Alternating current, or AC for short, changes direction all the time. Direct current is used in small devices like flashlights. Our homes are wired for alternating current, which is much more efficient than direct current. It is easy to change, or transform, to a higher or lower voltage. This makes it easier and cheaper to send over long distances.

A power plant produces a **current** of several thousand volts. A transformer then increases, or steps up, this current to several hundred thousand volts to send it over the high-voltage power lines. At the other end of the lines, the **voltage** is decreased, or stepped down, to several thousand volts. Near a house, a pole transformer decreases it even further, to about 110 volts.

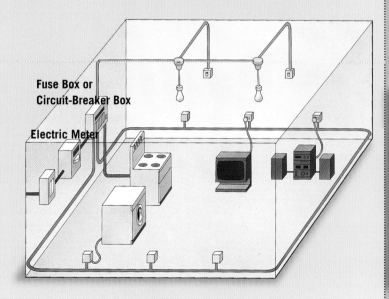

Fuse Box or Circuit-Breaker Box

Electric Meter

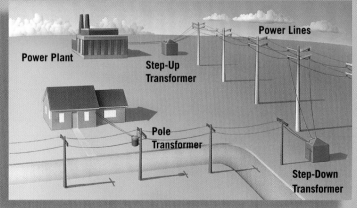

Power Plant

Step-Up Transformer

Power Lines

Pole Transformer

Step-Down Transformer

What Is Magnetism?

A magnet looks like an ordinary piece of metal, but it is surrounded by an invisible **field** of **force** that affects any magnetic material brought near it. A magnet can hold memos to a refrigerator door. Nails and paper clips magically cling to a magnet, and it can attract or repel another magnet.

The two ends of a magnet are called north-seeking and south-seeking poles and are labeled (N) and (S). When a magnet is suspended freely, it will line up north to south because the magnet's poles are attracted to the Earth's magnetic poles.

The first important use of the magnet was as a compass. The north-pointing needle of the compass allowed ships to stay on course even when they were out of sight of land. Today magnets come in many shapes, are made of a variety of materials, and have become a vital part of modern life.

Opposite Poles Attract

Like Poles Repel

Tiny iron filings demonstrate a basic law of magnetism: Opposite poles attract and like poles repel. When two magnet poles come close to each other, they exert either a pushing or a pulling force, as shown by the filings clinging to the north and south poles of these bar magnets. The red north pole attracts the blue south pole, but the two red north poles repel each other. Two south poles will repel each other as well.

Magnetite: Natural Magnet

Since ancient times, humans have been aware of the power of magnetism. The Chinese, Arabs, Greeks, and Romans discovered a type of black rock—actually iron **ore**—that had the magical power to attract and repel other pieces of metal. When hung freely, the same side of the rock would always point in the same direction. This natural magnetic rock is called magnetite. The Greeks found large deposits of this ore in a region called Magnesia, now a part of Turkey. The words "magnetite" and "magnet" probably came from this region's name. Magnetite is often referred to as a lodestone. "Lode" is an early English word meaning "to lead." Early compass needles were magnetized by using a piece of lodestone.

Making a Magnetic Field Visible

Place a bar magnet on a piece of paper covered with iron filings, and the magnet's invisible lines of force—or magnetic field—immediately appear. The diagram at right shows that the magnetic field is a continuous loop that runs from the magnet's north pole around to the south pole and back again. The field is strongest at the poles, where the lines of force converge.

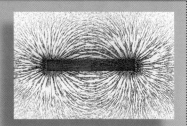

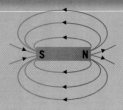

Characteristics of Magnets

Inside a metal object such as an ordinary horseshoe, there are magnetized regions called magnetic domains. These are arranged randomly, with their north and south poles facing in different directions. The forces cancel each other out and the object is unmagnetized. But in a horseshoe magnet, the domains are aligned, with all the north poles facing one direction and the south poles in the other. Magnetic poles always come in pairs, north and south. No matter how many times a magnet is cut in half, each piece will still have two poles because the magnetic domains remain aligned.

Magnetic Domain

N

N
S
N
S
N
S

S

Once a magnet, always a magnet.

Let's Compare

Permanent and Temporary Magnets

Permanent magnets are often made of iron, aluminum, nickel, and cobalt. A magnet stays permanent as long as its domains are aligned. If it is struck hard, the domains can be jolted out of alignment. Heat can also demagnetize a magnet.

Materials like paper clips can become temporary magnets during magnetic induction. The domains become slightly aligned, giving the clip a magnetic force. The magnetized clip then attracts the one below it, forming a chain. Remove the magnet, and the domains return to normal; the clip is no longer magnetized.

Magnets for Health

Since magnetite had such mysterious powers, even in ancient times people began to wonder if it could be used for healing. Starting in the 16th century, a variety of magnetic treatments were introduced. Ground magnetite mixed into a salve was said to have healing powers. In the 1780s people held iron rods sticking out of a vat of "magnetic water" while a magnetic wand was waved over them. But the "cures" were suspect.

Today magnets have been put to a more practical use. One diagnostic procedure is magnetic resonance imaging, or MRI. This system uses a strong but harmless magnetic field to penetrate the body's soft tissue without damaging it. A 3-D image of the internal organs is projected onto a computer monitor that shows any abnormalities, such as tumors or arthritic joints.

The Many Uses of Magnets

Magnets are found in all sorts of objects that we use every day. In addition to the common refrigerator magnet, there are numerous items that have hidden magnets. The magnetized black strip on the back of a credit card holds information that is read when it is passed through a credit card scanner. The plastic tape inside music and video cassettes is coated with iron oxide. These magnetized particles hold the information that is translated into sound and pictures. Computers store data as magnetic patterns on plastic disks that have a magnetized coating. Radio and television speakers produce their tones with a vibrating magnet. Doorbells and burglar alarms use magnets to activate the sound.

Magnetic Earth

Earth is actually a giant magnet, and just like a regular bar magnet, it has two magnetic poles. Surrounding Earth is a gigantic magnetic **field** stretching far beyond the atmosphere into space. This field protects Earth from the Sun's harmful solar radiation. Sometimes particles get caught in the magnetic field, exciting **atoms** and **molecules** in the atmosphere and producing a spectacular light show called the aurora *(right)*.

No one is absolutely sure what gives Earth its magnetic field. Scientists think that it is caused by the motion of Earth's **molten** inner core. They theorize that Earth's rotation generates an electric **current** that produces Earth's magnetic field. Astronomers now know that every large spinning body, such as a planet or a star, has a magnetic field.

Flip-Flopping Poles

Strange But TRUE!

About every 300,000 years, Earth's magnetic poles reverse themselves; that is, the north and south poles trade places. This is known as polar reversal and has probably happened ever since Earth was formed. How can scientists possibly know this? The clue lies in magnetic striped patterns of iron ore found in certain rocks. The magnetic particles in molten rock, such as in the lava flow at right, align themselves with Earth's magnetic poles. When the rock hardens, a permanent record of Earth's magnetism stays in the rocks. Although these magnetic stripes can't be seen, geologists can use special instruments in order to read them.

Geographic and Magnetic Poles

Earth has two sets of poles, the geographic and magnetic poles. They are not located in the same place. The geographic poles—the ones usually shown on maps—are located at the fixed points where Earth rotates on its axis.

The magnetic poles lie about 1,600 km (1,000 mi.) away from the geographic poles, at an angle known as the magnetic declination, or variation. The magnetic poles are not fixed. They are constantly wandering within the magnetic variation. The north magnetic pole (where a compass points) is located at present off Ellef Ringnes Island in the Arctic region of Canada. The south magnetic pole is located off the Adélie Coast in Antarctica.

Earth's immense magnetic field, seen as purple lines in the diagram below, extends far out into space. This region is called the magnetosphere, and it surrounds Earth with a protective shield, deflecting the Sun's constant bombardment of harmful radiation.

Charged particles of the Sun's radiation sometimes get trapped in Earth's magnetic field and spiral down near the poles. The particles hit atoms and molecules in the atmosphere, making them glow in different colors that pulsate across the night sky *(left)*. This spectacular show is usually visible only near the poles. In the Northern Hemisphere, it is called an aurora borealis, or northern lights. The same effect in the Southern Hemisphere is called an aurora australis, or southern lights.

Finding Magnetic North

A compass has a magnetized needle that rotates freely in a circle. It points to the north magnetic pole, not the north geographic pole. A compass needle can't point straight down, so how do scientists, pilots, or explorers know when they are directly over the pole? They use something called a dip needle, which is really a compass on its side. A dip needle points vertically instead of horizontally. When it is directly over the north magnetic pole, the dip needle points straight down. When it is over the south magnetic pole, the north end of the needle points straight up.

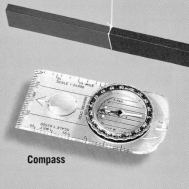

Compass

Dip Needle

Built-In Compass

Wearing a radio transmitter just a little bigger than a deck of cards, a sea turtle returns to the ocean. The transmitter sends signals up to a global weather satellite so scientists can track the turtle's migration from the United States to Japan. Some research indicates that some animal species, such as homing pigeons, sea turtles, and monarch butterflies, have a built-in compass that uses Earth's magnetic field to help them navigate over incredibly long distances.

North Geographic Pole

North Magnetic Pole

South Magnetic Pole

South Geographic Pole

Electromagnetism

Electricity and magnetism are closely related. Magnets can be used to produce electricity, and an electric **current** can be used to make a temporary magnet. This special link between electricity and magnetism is called electromagnetism.

A wire carrying an electric current always produces a magnetic **field.** A piece of coiled, insulated wire with a current flowing through it is called a solenoid. As long as current is passing through it, a solenoid acts just like a bar magnet, having a strong magnetic field with a north and a south pole.

A magnet created in this way is called an electromagnet (*opposite*). An electromagnet has many advantages over a permanent magnet. It can be turned on and off as needed, and the strength of the magnetic field can be varied by the amount of current flowing through it.

Current Off

Iron filings lie scattered around a wire that has no electric current flowing through it.

Current On

As soon as an electric current flows, the iron filings line up around the magnetic field that's produced.

Magnetic Field around an Electric Wire

When a magnetic field is generated by electric current in a wire, the magnetic field lines form circles around the wire. This effect can be seen in the picture above, in which an electric wire has been placed through the center of a horizontally placed piece of cardboard. With the current turned off, iron filings on the cardboard lie scattered randomly (*above, left*). When the current is turned on, it generates a magnetic field that aligns the iron filings in a circular pattern around the wire (*above, right*). The magnetic field exists only while the current is flowing through the wire. When the current stops, the magnetic field disappears.

French physicist André-Marie Ampère began his own experiments with electricity and magnetism and built on Oersted's findings. Ampère discovered that a magnetic field exists in a circular pattern around every wire carrying a current: The stronger the current, the stronger the magnetic field.

People Hans Christian Oersted

In 1820, Hans Christian Oersted, a Danish professor, was performing a classroom experiment. He noticed that an electric current flowing through a wire deflected a nearby compass needle so that it no longer pointed north. Oersted realized the current was producing a magnetic field. He was the first to find the connection between electricity and magnetism. The Oersted Medal (*right*) is now awarded to outstanding instructors in the field of physics by the American Association of Physics Teachers.

What's an Electromagnet?

An electromagnet is a type of magnet produced by an electric current. This kind of magnet is simply a coil of insulated wire, or solenoid, with an iron core. When the wires are attached to a flashlight battery, the current flowing through the coiled wire will create an electromagnet, such as the one at left picking up a mass of paper clips. Giant-sized electromagnets (*above, right*) are often used in junkyards. Since magnets attract only certain types of metal, such as iron and steel, these electromagnets are used for separating one kind of metal from another. They are also strong enough to lift several tons of scrap iron or steel girders. When the current is turned off, the scrap metal falls to the ground.

Maglev Train

Imagine a train that has no wheels and no engine but can move at speeds up to 480 km/h (300 mph). Developed in Japan and Germany, magnetic levitation (maglev) trains float on a cushion of air above a single rail. Electromagnets attached to the underside of the train and on the rail levitate the train and propel it forward smoothly along the track. Since a maglev does not touch the rail, there is little **resistance** or **friction.** This allows the train to use less **energy** while moving much faster and almost silently.

Let's Compare

Making Electricity with a Magnet

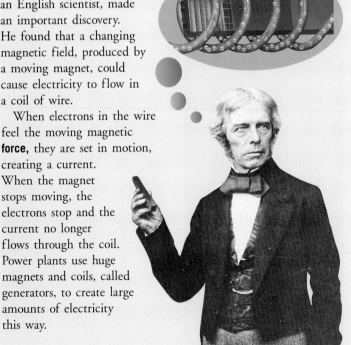

In 1831, Michael Faraday, an English scientist, made an important discovery. He found that a changing magnetic field, produced by a moving magnet, could cause electricity to flow in a coil of wire.

When electrons in the wire feel the moving magnetic **force,** they are set in motion, creating a current. When the magnet stops moving, the electrons stop and the current no longer flows through the coil. Power plants use huge magnets and coils, called generators, to create large amounts of electricity this way.

Generator vs. Motor

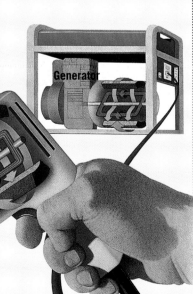

A generator and a motor both use a magnet and a coil of wire to change one type of energy to another type of energy. A generator converts mechanical energy, such as water power, coal, or gas, into electric energy. A motor is simply a generator in reverse. It changes electric energy into mechanical energy to do work, like the power for an electric drill or other appliance.

Generator

Motor

What Are Waves?

Toss a pebble into a pond and watch the ripples spread from the center in growing rings. These disturbances are waves, the same as the much larger ocean waves, or waves that carry sound, and even the microwaves that heat your leftovers.

Waves don't actually move water or sound forward as they advance. Instead, they transfer **energy** from one place to another in periodic—or rhythmic—cycles. Some types of waves, called mechanical waves, require a material, or **medium,** to travel through. Earthquakes, for example, are caused by destructive seismic waves that rumble through Earth's crust as they transfer their energy across great distances. Sunlight, by contrast, travels in **electromagnetic waves** that require no medium as they speed through empty space from the Sun to Earth and the rest of the solar system.

Wave Parts

One complete cycle of a wave, falling from crest to trough and rising to crest again, is called a **wavelength.** In a transverse wave—that is, a wave going across, such as a water wave—the amplitude is the distance of the water surface from its resting position to the crest and is a measure of its strength or size. Compressional waves must press against something to **propagate,** or travel. Sound waves travel this way, by pushing against nearby **molecules** in a rhythmic cycle of **compressions** and **rarefactions.**

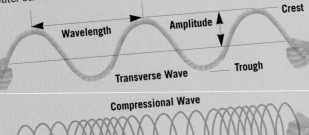

Wavelength · Amplitude · Crest · Trough

Transverse Wave

Compressional Wave

Catching a Wave

Surfers at Hawaii's Waimea Bay *(below)* ride a cresting ocean wave. A gradual slope at this famous beach allows waves to build up large amplitudes as they speed to the shore. Although it appears that the waves crash into the beach, it is their energy that is transmitted. The water moves mostly up and down as the wave passes.

Tsunami!!!

Tsunamis are giant ocean waves, powerful enough to topple structures such as the massive statues on Easter Island *(below).* They are triggered by earthquakes or volcanic activity on the ocean floor. A tsunami has an extralong wavelength whose amplitude can swell into a terrifying wall of water, surging at up to 800 km/h (500 mph).

What's Frequency?

When you turn the dial of a radio to your favorite station, you are tuning in to waves of a certain frequency. Frequency describes the number of wave crests that pass a given point in one second. Frequency is measured in units called hertz, named for Heinrich Hertz, the German scientist who discovered radio waves in 1886. One hertz equals one wave per second. A radio station with a frequency of 92.3 megahertz, for example, transmits 92,300,000 waves per second.

Try it!

Achieve Resonance

If you pump your arms and legs a certain way on a swing, you will discover after a while that you no longer have to work as hard to stay in motion. At this point, you swing at the natural frequency of the swing, and very little effort is required to maintain the amplitude of the **vibration** you created. When you pump the swing at its natural frequency, you create a **resonance.** Most objects, including playground swings and bridges, have some degree of natural elasticity that allows them to vibrate at some frequency.

Wave Behavior

Reflection

When waves encounter a new material, they bounce back, or they may bend. Light waves bounce, or reflect, when they hit a flat, mirrorlike surface, creating an image. Sound waves reflect by bouncing off walls, ceilings, and distant mountains, sometimes creating echoes. Waves can also bend, or **refract,** as when ocean waves bend around a sand bar or light passes into water, which slows its speed.

If two waves crash into each other, they create **interference,** resulting in a change. A rogue wave in the ocean, for example, is an unusually large wave that occurs when waves meet in the same portion of their phase and double their amplitude. If one wave's crest encounters the trough of another, they can cancel each other out, and the water flattens. Sound wave interference is used in car mufflers to reduce engine noise.

Placid water in Washington State's Rainier National Park creates a natural mirror, reflecting the snowcapped peak of Mount Rainier.

Hello ... Hello ... Hello

You can hear an echo whenever sound waves **reflect** from a solid barrier, such as a mountain or a brick wall. The wall forces the waves to travel back toward their source. If you clap your hands some 15 m (50 ft.) away from a wall, reflected waves will return after the original source waves have stopped. You may be able to distinguish the reflected sound from the original sound and hear an echo.

IT'S THE LAW

Law of Reflection

A light wave that bounces back from a reflective surface follows the law of reflection: The angle of incidence, or in-coming wave, equals the angle of reflection. The reflected image, called a virtual image, will look like the original image in reverse. Distortions in the reflective surface, such as curves or texture, will affect the way the light is reflected.

Plane Mirror

Reflected Angle

Incident Angle

Virtual Image

Object

Refraction in Action

The flower stem in the vase at right looks broken because light waves are bent, or **refracted,** as they pass from one substance to another. The greater the difference in the **densities** of the two substances, the more the light bends. In 1621 Dutch astronomer Willebrord Snell showed that every substance bends light differently, a quality he called its refractive index. A diamond, a very dense substance, has a high refractive index of 2.42. Air is much less dense and has a low index of 1.0. The refractive index of water is 1.33.

The Nature of Light

When white light passes through a **prism** *(below),* the light breaks into a band of colors, beginning with red and followed by orange, yellow, green, blue, and violet. You have seen this phenomenon in the colors of the rainbow. Each color has a different wavelength, with crests and troughs. Violet has the shortest wavelength and bends the most. Red, with the longest wavelength, bends the least.

Interference

The swirling colors in a soap bubble are caused by the way light reflects off its surface. When light strikes the bubble, some light waves bounce back immediately; others penetrate the soap film and bend and **reflect** out of step from the other light waves. When they do not match up, they weaken each other, causing interference, and the light breaks into the colors of the rainbow. Where the crests match up, the waves strengthen each other. Where the trough of one wave meets with the crest of another, the waves cancel each other out, resulting in black spots.

Reflected Rays

Incoming Rays

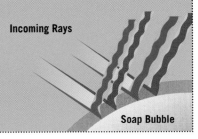

Soap Bubble

Diffraction

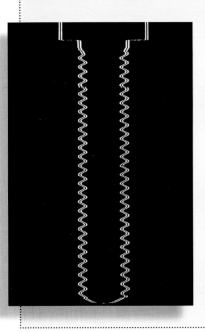

Diffraction describes a wave's ability to spread out, or to bend around corners. You have experienced diffraction if you have ever heard music coming from another room. The sound waves had to bend around walls and doorways to reach your ears. The photo at left, taken in a laboratory through a special filter, shows how light is diffracted around the edges of a common bolt from the hardware store.

Sound Waves

Every sound you hear, from a dog's bark to the drumbeat of your favorite band, is caused by **vibrations.** The vibrations travel to your ear in a series of **compressions** and **rarefactions** called sound waves. **Energy** from the vibrations stirs nearby air **molecules,** allowing the waves to **propagate,** or spread out, as they move away from their source.

Both the material a sound wave passes through and the temperature of that material affect the speed of sound. Sound travels about five times as fast through water as through air at room temperature, and about 20 times faster through iron or steel. Sound also travels faster on warm days than on cold ones, because cold slows the motion of air molecules. In outer space, where there are no molecules for sound waves to push against, astronauts must communicate by radio, even if they are standing right next to each other.

What's a Sonic Boom?

A cloud of condensed water vapor shows up behind a U.S. Navy Hornet fighter as it breaks through the sound barrier. When jets travel faster than the speed of sound—about 1,184 km/h (740 mph), depending on air temperature—they compress air molecules at the front and sides of the plane. The molecules then collide with the relatively still surrounding air, creating a tremendously loud shock wave, called a sonic boom, that can crack windows and shake your home. A photographer could take this rare photo because the amount of moisture in the air was unusually great, which caused the vibrations to show up as a cloud.

Surrounded by Sound

Seiji Ozawa Hall at Tanglewood Music Center in Massachusetts has been specially built for the best **acoustics** possible. The ceiling, walls, and furniture —all made of wood—allow sound waves to spread out evenly as they propagate away from the stage. Heavy carpets and curtains absorb unwanted sound. In poorly designed halls, sound waves can interfere with each other, creating unusually quiet or loud spots.

Loud!

Loudness, also defined as a sound's intensity, depends on the amount of **energy** a sound wave carries and is measured in decibels. A whisper registers at about 20 decibels, whereas a rock concert can exceed 120 decibels, enough to cause permanent hearing loss. A shrill sound, such as a whistle, is high-pitched, which means it has a high frequency, but low rumbles from thunder have a low frequency and a low pitch.

Doppler Effect

When a fire engine advances, the sound of its siren rises in pitch and falls as it passes by. This change in sound is called the Doppler effect. As the vehicle comes closer, the sound waves pile up ever faster in front *(below)*. The frequency of the sound waves increases and their **wavelengths** get shorter, resulting in a higher-pitched sound. As the vehicle passes, the wavelengths get longer, and the wailing sound of the siren drops to a lower-pitched sound.

Seeing with Sound

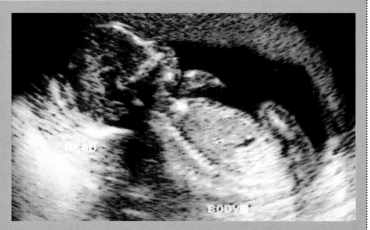

An expectant mother can view the progress of her baby's development on the monitor of an ultrasound machine *(above)*. Ultrasound creates a picture of liquid-based objects inside the body by directing high-frequency sound waves at an area and measuring the time it takes for the waves to reflect back. Based on the data, the computer builds a picture on the monitor, providing parents and doctors with important information about a baby's growth.

Shattering Sounds

Nothing more than a sound wave caused this glass to explode *(right)*. Vibrations from sound, such as a continuous, high-pitched note, can cause sympathetic vibrations in nearby objects, producing a **resonance.** If the frequency of the vibrations matches the natural frequency of a fragile object such as a glass, the amplitude of the waves may build up high enough to shatter the glass.

Whale Songs

Grunts, whistles, chirps, and squeaks make up the songs of humpback whales. (A humpback is shown above, right near a researcher's microphone.) Scientists disagree on the purpose of the songs. The whales may sing to attract mates, declare social status, or send other kinds of messages. The low-frequency sound waves of their rumblings carry for hundreds of miles through the water and may be detectable across entire oceans.

What Is Light?

Light is a form of **electromagnetic wave** that we can see, among a range of invisible **wavelengths**. All of the light and colors we see represent just a tiny fraction of the total frequency range of the electromagnetic spectrum. The electromagnetic waves can be arranged on a spectrum—a kind of graph—that ranks them from the longest radio waves to the shortest gamma rays *(right)*. Light requires no **medium,** but travels in oscillating electric and magnetic fields. Although all of the waves travel at the same **speed**—the speed of light—each group has a different wavelength and carries a different amount of **energy.** The Sun is the main source of Earth's electromagnetic **radiation,** emitting all wavelengths of light. Not all of those wavelengths reach Earth, however. Ozone in the atmosphere absorbs a lot of ultraviolet radiation, primarily allowing wavelengths of infrared and visible light to reach Earth.

People | William Herschel

British astronomer William Herschel accidentally discovered infrared radiation in 1800. He was searching for a colored telescope filter that would allow him to look into the Sun safely to view sunspots. Herschel's discovery was the first evidence of electromagnetic waves that can be felt but not seen by human eyes.

The Electromagnetic Spectrum

Gamma Rays

Gamma rays have the shortest wavelength, and their **photons** carry an enormous amount of energy. They are emitted in **nuclear** reactions and explosions, seriously damaging any living cells they pass through.

X-Rays

X-ray photons easily penetrate soft substances such as skin and cloth, but slow as they pass through bone or metal. As a result, they are useful both to doctors examining bones and to airport security guards, who scan luggage for hidden weapons.

Radio

The longest waves in the electromagnetic spectrum are low-frequency radio waves, which are used to broadcast radio, television, and cellular phone signals.

Microwaves

Microwaves cook food quickly by heating the moisture present in food. They are the shortest of radio waves and are also used to transmit radar signals.

Infrared

All objects give off infrared rays. A thermograph *(left)* is a special photograph taken of heat, indicating the various temperatures of an object, which is usually cooler on the outside. Many restaurants use infrared light to keep food warm.

Visible Light

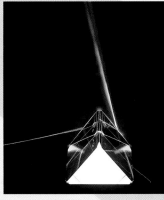

A **prism** separates the white of visible light into a rainbow of colors. Most people see five true colors: red, green, yellow, blue, and violet. When mixed together these colors form an infinite range of hues.

Ultraviolet

Ultraviolet photons are energetic enough to burn skin and damage eyes. Although Earth's ozone layer offers some protection from the full force of ultraviolet radiation, sunglasses and sunscreen are still needed for anyone planning a day at the beach.

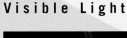

How **Fast?**

Catching Up to Light

The universe has a speed limit: 297,600 km (186,000 mi.) per second—the speed of light. Humans can't travel at this speed, but the concept of a light-year—the distance light travels in one year—helps scientists determine the relative distances of stars and planets. Traveling at the speed of light, it would take about 1.3 seconds to reach the Moon, 8.5 minutes to reach the Sun, and 30,000 years to get to the center of the Milky Way.

Applications of Light

For millions of years, plants have used the **energy** carried by sunlight to make the energy necessary for life. Without this vital process, called photosynthesis, life on Earth would not exist. Humans have only recently started putting light to use. In 1905, Albert Einstein described how particles of light energy, or **photons,** separate an **electric charge** in **atoms,** particularly on some types of sensitive metals. Known as the photoelectric effect, this phenomenon stimulates film to record images and powers solar cells in calculators. Another light source, invented in 1960, is a laser, which generates a concentrated beam of photons of the same frequency. This narrow beam can transmit large amounts of energy strong enough to cut through stone and metal. Lasers also perform delicate eye surgery, aid in precision measurements, weld giant pieces of steel, operate automatic flush toilets, remove tattoos, and carry as many as 450,000 simultaneous telephone conversations over a cable of optical fibers.

Wave or Particle?

Some properties of light can only be explained if light is a wave. Like other waves, light can be **reflected** and **refracted,** and it has a measurable frequency, as is demonstrated by the different **wavelengths** of the electromagnetic spectrum *(pages 80-81).*

Other properties of light can be explained only if light behaves like a particle, or photon. The photoelectric effect creates an electric current when photons strike the surface of certain light-sensitive metals. Photon energy increases as frequency increases and the wavelength decreases. In fact, light behaves like both a wave and a particle. This is called the duality of light.

Would **You** *Believe?*

Solar Car

Aurora I, winner of the 1999 World Solar Challenge race from Darwin to Adelaide, Australia, sits in Adelaide's rush-hour traffic. The solar car completed the 3,000 km (1,875 mi.) race in 41 hours and 26 minutes. Photo-voltaic cells on the roof convert sunlight into electricity that helps the car go up to 120 km/h (75 mph). Solar power offers solutions for many of the world's energy needs. In one hour, the Sun releases more energy onto the Earth than all the world's people consume as **fuel** in one year.

Supermarket Science

At the supermarket, a beam from a helium-neon laser reads product information embedded in the black lines of a Universal Product Code. This optical information is converted to electric pulses and transmitted to the register. At the register, the product name and price pop up on the screen and update store inventory as the item is rung up. Lasers stimulate huge numbers of photons of identical frequency to generate their beams. The information they can carry makes them a valuable tool for transmitting data and communications. Depending on the materials the laser is made from, it can generate light at wavelengths ranging from infrared to ultraviolet.

What's a Laser?

A flashlight *(below, left)* casts a spreading beam of white light that is made up of wavelengths of every color in the spectrum. A laser—the term stands for "lightwave amplification by stimulated emission of radiation"—emits a powerful beam of "coherent" light of identical, parallel wavelengths.

Listening with Light

You probably own your very own laser—hidden inside the workings of a compact disk player. More than 4.8 km (3 mi.) of spiraling microscopic grooves and flat spaces etched into the reflective side of a compact disk serve as a map for the laser's beam. The beam bounces off a mirror and flashes through a focusing lens to the underside of the disk *(diagram, right)*. The laser reads the disk's markings as binary numbers (0s and 1s) and converts them to electronic codes, reproducing the sound as it was recorded in the studio.

Compact Disk

Laser Beam

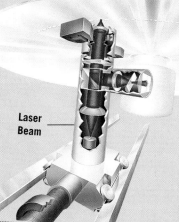

Fiber Optics

Optical fibers, made of strands of silicon glass up to ¹⁄₁₀ the thickness of a human hair, transmit digital information in the form of laser pulses. Thousands of telephone conversations or other pulses of information can be transmitted at the same moment through their cables.

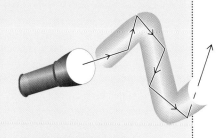

Total internal reflection *(illustrated above)* directs laser pulses through strands of fiberoptic cables. "Repeaters" placed about every 48 km (30 mi.) along cables ensure that the signal stays strong as it passes over great distances.

What Is Color?

White light from a sunbeam or a light bulb contains all the colors of the rainbow. Isaac Newton discovered this in 1665 while experimenting with a glass **prism,** observing that a beam of light that passed through the glass always "split" into the same band of red, orange, yellow, green, blue, and violet. When passed through a second prism, the colors recombined and made white light.

Newton deduced that light is made up of all colors, each with its own **wavelength** that is **refracted** to a different angle by the sharp edge of the prism. Newton further deduced that color depends on reflection. An apple appears red because it absorbs all the colors of the spectrum except red, which it **reflects** back. If a blue light shines on an apple, the apple appears black, because it absorbs all blue wavelengths of light, leaving no colors to reflect.

Rainbow

Just after a rain shower, water droplets in the air act like a million tiny prisms, refracting sunlight into the familiar colors of the rainbow *(below)*. Sometimes light reflects twice in the water drops and produces a secondary rainbow with the colors ordered in reverse.

Primary Colors

Red, blue, and green make up the primary colors. If they are projected as light, the colors —called additive primary colors—will combine as white. Other combinations will produce all of the other colors in the spectrum. When two primary colors combine, they make one of the secondary colors: cyan, yellow, or magenta. Dots of primary-colored light make up the pictures on television screens *(right)*. Lighting up different portions of the strips will reproduce most of the other colors in the spectrum.

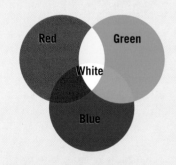

Secondary Colors

Objects that do not produce light on their own are colored by subtraction. Some wavelengths of light falling on an object will be absorbed, while its true color will be reflected back. Combinations of cyan, magenta, and yellow in pigments can make all of the colors of the spectrum. In color printing, a page like the one below goes through a four-color process, during which it is

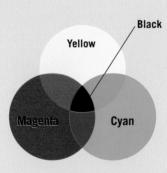

first coated with black ink, followed by layers of cyan, magenta, and yellow to create the finished picture.

Fogbow

A curious gray fogbow hovers over the Antarctic landscape shown above. When water drops in the atmosphere are as small as those found in clouds or fog, diffraction of light passing through the drops spreads each color of the spectrum out into a wider angular pattern. The result is that the colors overlap, producing a broad, white bow.

Sometimes on a rainy night you can also see a moonbow, which contains all of the colors of the rainbow but appears gray. Color sensors in your eyes cannot distinguish the different shades in the dim light.

Colors of Nature

Photosynthesis ➤

Trees and plants reflect lush shades of green, a sure sign that photosynthesis is going on. Pigments called chlorophyll are responsible for reflecting the green color. The pigments absorb sunlight and combine it with water and carbon dioxide to make plants grow. In autumn, plants lose their summer green when they stop producing chlorophyll to prepare for winter.

Sending Signals ➤

The hibiscus wears a brilliant red, a color meant to attract passing hummingbirds. The birds sip the flower's nectar and pick up bits of pollen to be passed to the next flower, ensuring next year's growth.

Scattering Light ➤

Startling shades of blue in this iceberg are the result of ice crystals absorbing red light and scattering blue more than red; this is the same principle that explains the color of the sky. The green sea below the iceberg is colored by phytoplankton, whose chlorophyll reflects green.

What Are Optics?

L enses help us see things more clearly by bending, or refracting, light. Human eyes contain natural lenses that focus on images by tightening and relaxing the muscles in the eye.

Man-made lenses of glass or plastic cannot adjust as easily. They possess a fixed **focal length** where light bends as it passes through.

In a convex lens—a lens that is thickest in the middle—light **converges**—that is, the parallel light waves are brought together at a fixed point, making objects appear larger.

In a concave lens—a lens that is thinnest in the middle—light **diverges,** or spreads out, making objects look smaller. The image seen through a concave lens is always smaller than the object being viewed. Lenses are important in a variety of optical instruments, ranging from microscopes to telescopes, that manipulate light to examine all kinds of things, from very small to very large, nearby to faraway.

Up Close and Personal

A huge eye stares through the convex lens of a magnifying glass at left. Light rays passing through the lens bend to meet at a focal point, located about 12.5 cm (5 in.) from the lens. All objects viewed between the lens and the focal point lie within the focal length, and they will look greatly magnified.

How Far Can We See?

T he view through the Hubble Space Telescope *(right)* shows spiral galaxy NGC 5194, one of the most distant objects in the universe, lying more than 13 million light-years from Earth. It would take us 13 million years of travel at the speed of light to reach this galaxy.

The salmonella bacterium *(right),* which is shown enlarged thousands of times, is so small it can be seen only through a scanning tunneling microscope.

Convex Lens

Light passing through a convex lens *(below)* bends inward to a focal point determined by the thickness of the glass and the curvature of the lens. Convex lenses are used to correct far-sighted vision. They pull images closer to focus at the retina and help as reading glasses.

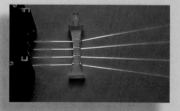

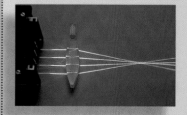

Concave Lens

A concave lens *(above)* spreads light rays out so that they cannot meet. The eye sees a smaller image through the lens, as the lens projects images farther back in the eye to focus on the retina. These lenses are used to help people who are nearsighted see faraway objects.

Inside a Camera

Focusing on a furry friend, a photographer adjusts the distance between the lens and the film in his camera to make sure he gets a clear picture. Light rays pass through a camera's focusing lens, producing an upside-down image on the film *(diagram, right),* which is located at the focal point of the

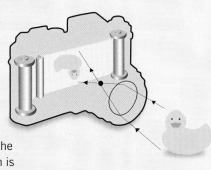

lens. Light-sensitive film placed at the rear of the camera records the image. Too much light will cause the final photograph to look too pale, whereas too little light will make the photo look too dark.

Seeing in the Dark

Illuminated through the eerie green light of night-vision goggles, a criminal is captured in a city subway system. Night-vision goggles are sensitive to a broad range of light, from visible through infrared. Light comes through the goggles, **reflects** off an image intensifier, and is converted to an electric charge

that is, in turn, converted to green light, allowing the viewer to remain invisible as he or she explores the dark.

Curved Mirrors

Light sources placed at the focal point of concave mirrors, as in flashlights, car headlights, and the lighthouse at right, are reflected in strong beams of parallel light.

Convex mirrors curve outward *(below)* and let light rays bend away from the mirrors' surface. They produce a smaller image but allow large areas to be

viewed. Because they show a wider field of view than flat mirrors, convex mirrors are used as rearview mirrors in cars and can help prevent accidents by signaling dangerous sites, like this treacherous curve in Italy.

Try it!

Watch Television Upside Down!

To watch your favorite television show upside down, all you need is a magnifying glass, a large piece of cardboard or paper, and, of course, a television. Turn off all of the lights in the room and stand about 3 m (10 ft.) from the set. Position the lens about 15 cm (6 in.) from the paper and move the paper back and forth until the image

comes into view. Light waves from the television screen travel through the magnifying glass from beyond the focal length, which causes them to converge and cross, like light rays through a camera lens, making the image you see projected on the paper inverted and smaller than the picture on the screen.

Can You Believe Your Eyes?

Have you ever noticed that the Moon appears larger when it rises near the horizon? The Moon is not any closer to Earth then, but your brain compares it with the nearby trees and houses rather than the far-off sky, creating the illusion of a larger Moon.

Illusions like this are the result of a partnership between your eyes and your brain. Your eyes collect light to see, and your brain interprets the information, using tricks and shortcuts to speedily process the millions of signals it receives. In everyday life, visual clues abound to help your brain fill in information about the size and shape of things, but sometimes you can be fooled. Most of the man-made illusions on these pages are possible because of the two-dimensional flat surface they appear on. Viewing such illusions helps us understand how the eyes and brain work together to create vision.

Which Way Is Up?

Where are these worm-like creatures marching to? Turn this book upside down, or sideways, and look again. It all depends on your point of view. Dutch graphic artist M. C. Escher was fascinated by mathematical principles of geometry, symmetry, perspective, and the concept of infinity, as seen in the seemingly endless cascade of staircases and creatures.

True Perspective

First perfected in Italy during the Renaissance, perspective creates the illusion of three dimensions on a flat surface. Artists call the viewer's attention to a single point on the horizon line of the painting. Objects approaching that point are rendered smaller than those in the foreground, mimicking the look of distant details in real life.

Making Movie Magic

Han Solo's spaceship, the *Millennium Falcon* from the Star Wars trilogy, looks like a toy next to one of its creators *(top right)*. Though only about 1.8 m (6 ft.) long, the model appears 10 times larger when viewed against the scaled-down background of the Cloud City *(bottom right)*. Scale is one of the most common special effects used in movies. In the Cloud City sequence, the background and ship were painted on glass with a transparent opening through which live action could be filmed, eliminating the need to create a full-scale model or build scenery.

Disguise

A clever master of disguise, the African flower mantis waits motionless *(above)* for unsuspecting prey. This type of camouflage, called mimicry, also helps to fool the insect's predators, who mistake it for a pretty but unappetizing flower.

Strange But TRUE!

Mirage

Mirages appear when light rays are refracted as they pass from a layer of dense, cool air to a warmer, less dense layer near the ground. From a distance, an observer sees an inverted image of trees, animals, and sky that seems to float just above the ground. Bent light rays from the sky can appear as a shimmering lake in the distance.

Try it!

With no distinct foreground or background in the image below, the shapes may "pop" in or out of perception. What images do you see? Are you looking at a vase or at two faces?

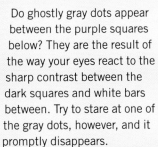

Do ghostly gray dots appear between the purple squares below? They are the result of the way your eyes react to the sharp contrast between the dark squares and white bars between. Try to stare at one of the gray dots, however, and it promptly disappears.

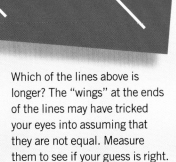

Which of the lines above is longer? The "wings" at the ends of the lines may have tricked your eyes into assuming that they are not equal. Measure them to see if your guess is right.

Which cat is largest? Because of the way the lines of perspective are drawn, your brain assumes that the cat in the "distance" must be largest, based on its size relative to its surroundings. Even when you know the cats are the same size, your brain continues to see the illusion.

Special Relativity

n the late 19th century, physicists were puzzled by a strange problem: Why is the **speed** of light always the same when we measure it, no matter whether the light source is coming toward you or is going away from you, or if you're speeding toward it *(below, right)*? This isn't true for other measurable things, like the speed of sound waves or the speed of moving particles.

Working at home, the young Swiss patent office worker Albert Einstein developed theories of space, time, **mass,** gravitation, and motion that solved this problem. His paper on the special theory of relativity, published in 1905, refers to special motion effects, such as the speed of light. The general theory of relativity, which followed in 1915, is largely about the nature of **gravity.** These new theories changed the way we understand the universe.

In 1931 Albert A. Michelson *(above)* flashed a beam of light through a 1.6 km (1 mi.)-long vacuum tube in Santa Ana, California, to determine the speed of light at 299,773 km (186,271 mi.) per second, now refined to 299,792 km (186,282 mi.) per second.

People

Albert Einstein

Albert Einstein *(left, writing an equation for relativity)* was quietly working in the Swiss patent office when he published his revolutionary special theory of relativity at the age of 26. He was soon recognized as one of the greatest scientists in the world, and in 1921 he was awarded the Nobel Prize for his theories. After teaching at universities in Germany, Einstein came to the United States in 1933 to escape the Nazi regime. He spent the rest of his life as an honored professor at Princeton University.

Observations Are Relative

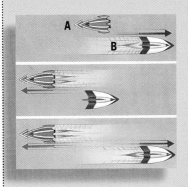

Relativity says that two people in different places (in different frames of reference) can see the same thing in different ways—and they will both be correct. For instance, observers on spaceships A and B *(left, top)* can see B as moving, A as moving, or both moving. Each observation is relative to a frame of reference.

A passenger tossing a ball on a speeding train sees it going straight up and down. But to someone seeing the train speed by, it looks as if the ball is moving in an arc. And each observation is correct!

Time Travel

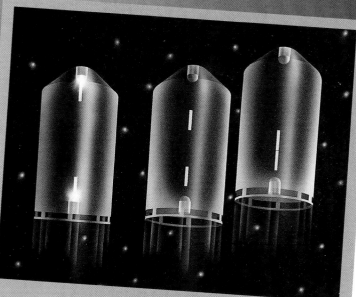

Einstein showed that speed would affect the way time passes. If a spaceship were to move near the speed of light, for example, time would pass more slowly on the ship than on Earth. Here's how: Two atomic clocks at either end of a spaceship send out light flashes to each other. The flashes begin together, but the lower flash lags as the ship accelerates. An outsider sees the ship's midpoint accelerate away from the lower end. But to onboard observers, the lower clock runs slow. Thus passengers would age more slowly than earthbound people.

Time Is Relative

Observer sees flashes together.

Observer moving to right sees right flash first.

Observer moving to left decides that right flash was first.

Einstein held that the order of events could be experienced differently if observers were in motion relative to one another —and each observation would be correct. Suppose, he said, that a person standing still sees two lightning flashes at the same time, left and right. A passenger on a train moving right sees the right flash first. If he imagines that the ground is moving past the train, but the train itself is still, then he believes that the right-hand flash happened first. If he convinces the standing observer that he's moving left relative to the train, the observer will agree with him that the flash at right was first.

Famous Equation—$E=mc^2$

Men on a naval **nuclear** aircraft carrier stand in formation to spell out Einstein's famous equation. He wrote the equation as a footnote to his special theory of relativity. It means that energy (E) equals mass (m) times the speed of light squared (c^2). The factor of c^2 (the square of the speed of light, a very large number) means that a small amount of mass converts into a very large amount of energy. Many years later this hypothesis was verified when the process of nuclear fission— the atomic bomb *(page 14)*—

was shown to release a vast amount of energy. Nuclear fusion *(page 15)* creates even more energy.

The equation also describes the energy obtained in a nuclear reactor, used to generate the power on an aircraft carrier like the one shown here.

For more than 200 years, Newton's idea of gravity *(pages 46-47)* as the attraction of all objects to each other seemed to work just fine. Newton's formulas predicted the return of Halley's comet; they were also important in the discovery of the planets Uranus and Neptune.

Then, in 1915, Einstein proposed his general theory of relativity. He expanded the concept of gravitational **force** by suggesting that there is a force between massive objects and light. Because of this force, light rays bend, leading scientists to conclude that space curves near large stars. Light rays that previously defined straight lines become curved. As a result, clumps of matter—such as stars or planets—make bumps and dips in space like heavy rocks in a soft sheet of rubber.

Fortunately, Newton's idea of gravity and Einstein's theory don't seriously conflict. Newtonian **mechanics** is good enough for almost all everyday science. Einstein's general relativity can mainly be seen with gravity on a very large scale. The big bang and the evolution of space both depend on the laws of general relativity.

In his general theory, Einstein also suggested that gravity and **acceleration** were equivalent—that their effects were the same. As an example, Einstein described an imaginary passenger in a closed spaceship *(right)*. That passenger would be weightless while traveling

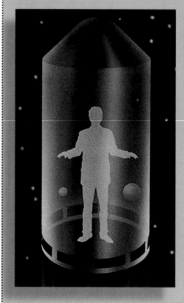

through space. If the ship accelerated suddenly *(left)*, he would be pressed to the floor. Unable to look out a window, the passenger would not know whether he was back on Earth, feeling its gravity, or accelerating through space. Because gravity and acceleration are equivalent, the sensations would be the same.

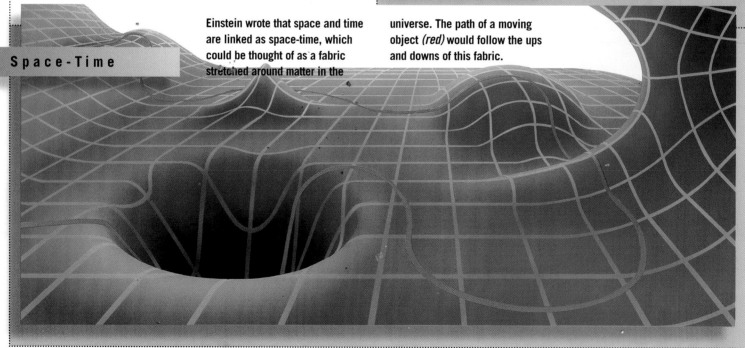

Space-Time

Einstein wrote that space and time are linked as space-time, which could be thought of as a fabric stretched around matter in the universe. The path of a moving object *(red)* would follow the ups and downs of this fabric.

Proofs of General Relativity

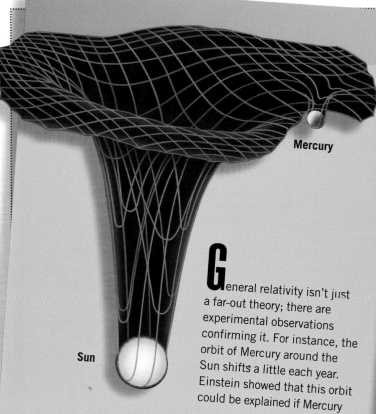

Mercury

Sun

was circling around the rim of a gravity well formed by the Sun. Even light follows the curves of space-time. Astronomers have seen that starlight that reaches Earth without passing close to the Sun will follow a straight line *(dotted line, below)*. When the Sun gets in the way, that light *(solid line)* can be seen to curve around the Sun's **mass.**

General relativity isn't just a far-out theory; there are experimental observations confirming it. For instance, the orbit of Mercury around the Sun shifts a little each year. Einstein showed that this orbit could be explained if Mercury

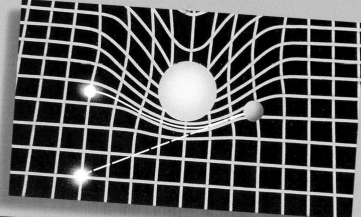

Using Relativity

Relativity affects many of the everyday devices that get signals from space. Earth-orbiting satellites *(right)* control the transmission of sound, print, and video messages. Their timing depends on atomic clocks. Because of the effects of gravity, however, these clocks will run differently from clocks on Earth, depending on where the satellites are relative to Earth's surface.

Today's ships, planes, and defense systems rely on signals from the GPS—Global Positioning System—to navigate. The hand-held GPS unit shown below includes the equations of general relativity in its calculations as it picks up satellite signals to pinpoint exactly where you are on Earth, and to guide you to anyplace else.

Physics Today and Tomorrow

I n the 1960s, physicists discovered that the **protons** and **neutrons** inside an **atom** *(pages 8-9)* are made of even smaller particles, called quarks. They identified six different types of quarks, called up, down, charm, strange, top, and bottom. In earlier studies physicists had discovered dozens of short-lived "elementary particles." Particles are affected by four forces—gravitational; electromagnetic; strong, or nuclear; and a force related to the nuclear force called the weak force—that hold everything in the universe together.

One of the main goals of physics at the beginning of the 21st century is to develop a single general theory that explains all of physics. Such a theory would include gravitation, electricity and magnetism, relativity (things moving at speeds close to the speed of light), and quantum **mechanics** (the physics of very small things such as atoms, protons, neutrons, **electrons,** and other elementary particles).

Quantum Physics

The quirky rules of quantum mechanics, describing **subatomic particles** and the forces that control them, dominated 20th-century physics. But the theory has practical results, too: transistors and lasers developed from quantum experiments. Once hard to find, tiny particles can now be analyzed in detectors such as the Stanford Linear Accelerator Center in California *(above, right).* The trace of a collision of particles appears at right.

Particles

Symbols representing subatomic particles drift across the universe below. A few of the exotic species are shown as symbols at right.

Quarks combine into protons and neutrons. Leptons are a class including electrons and neutrinos. Gluons "glue" quarks together. Bosons are particles that carry force. **Photons** carry electromagnetic force, whereas **gravity** operates through gravitons. Higgs bosons and leptoquarks, now extinct, were active in the early universe. An antiparticle has properties opposite from those of its counterpart.

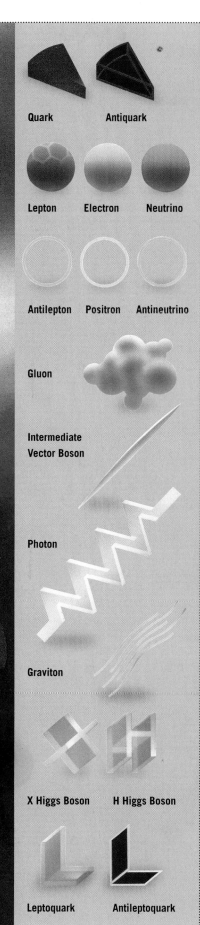

Quark   **Antiquark**

Lepton   **Electron**   **Neutrino**

Antilepton   **Positron**   **Antineutrino**

Gluon

**Intermediate
Vector Boson**

Photon

Graviton

X Higgs Boson   **H Higgs Boson**

Leptoquark   **Antileptoquark**

Black Holes

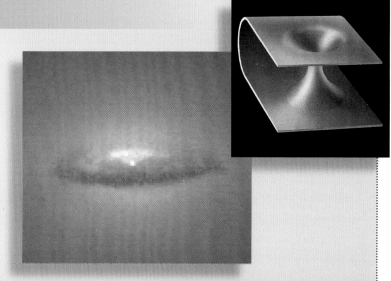

This colorized image of galaxy NGC 1097 may show **energy** from a black hole: an infinitely tiny, dense clump of matter with enormous gravitational pull. Some physicists suggest that black holes from separate universes once fused into tunnels called "wormholes" *(inset),* forming secret passages through space-time.

Uncertainty Principle

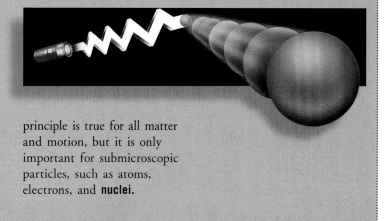

One of the first rules of particle physics, the uncertainty principle states that it is impossible to measure related properties precisely, because the very act of measuring things affects them. In this example *(right),* light from a flashlight locates an electron *(blue sphere).* However, as soon as the light's photons hit the electron, they change its **velocity** and position. The principle is true for all matter and motion, but it is only important for submicroscopic particles, such as atoms, electrons, and **nuclei.**

Cosmic Strings and Superstrings

How did the early universe get its shape? How did matter and energy form? Maybe from extinct cosmic strings *(illustrated at right).* In theory, these infinitely long, looping masses would have vibrated wildly, causing gravity waves. Shorter strings pulled matter around them into galaxies before vibrating out of existence. The tiny loops of superstrings may describe properties of and relationships between a variety of fundamental particles.

Natural Building Materials

Look at all the objects around you. How many different materials are these things made from? In our modern world we use thousands of materials, those that are found in nature and others, called **synthetic,** that have been manufactured.

Early humans first made things from natural materials found in trees, plants, animals, and the Earth's elements. Wood could be shaped into tools or shelters and burned for heat and cooking food. Plant fibers were woven into fabric to make clothing. Animal hair and skins—wool, leather, and fur—provided garments for warmth and protection. And the land itself supplied stone and clay for making weapons and pottery. At one time, the world's **natural resources** must have seemed limitless. Now we know that they are not, so we must use them with care.

Hardwood and Softwood

Wood can be divided into two main types: hardwood and softwood. Hardwood comes from deciduous trees, which means they shed their thin, flat leaves in winter. Baseball bats are made from the dense, springy hardwood of the ash tree *(top right).* Softwood comes from coniferous, or evergreen, trees, which have cones and hard, spiky needles. Violins are made partly from the light, easily shaped softwood of the spruce tree *(bottom right).*

Stone Sculpture

The world's largest sculpture, a memorial to the great Oglala Sioux war chief Crazy Horse, is being carved from a mountain of granite—one of the hardest stones—in the Black Hills of South Dakota. Begun in 1948, it is still decades away from being completed. The scale model below shows what the 172 m (563 ft.)-tall figures— eight times the height of the faces on Mount Rushmore— will look like when finished. The horse's nostril alone will be bigger than a house!

Fabulous Fibers

Linen

Linen is made from the inner bark of the flax plant *(right)*, which has long, thin, very tough strands like celery. Linen is one of the first textiles created by humans. This fabric does not tear easily and is cool to wear, but it is expensive to produce.

Silk

Silk comes from the protective cocoon spun by the silkworm moth *(left)*. Each cocoon may contain up to 900 m (3,000 ft.) of thread. Silk is the strongest natural fiber. It is also light-weight and soft and can be dyed in rich colors.

Cotton

Strong, inexpensive, and easy to clean, cotton is the best natural fiber for making clothing. The cotton plant produces rounded seedpods called bolls that burst open when ripe to reveal fluffy white cotton fibers *(right)*.

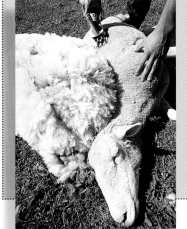

Wool

Wool is the soft, wavy hair of sheep, goats, camels, and alpacas. After being sheared, or cut, by machine or by hand *(left)*, the wool is cleaned, spun into threads, and woven into a fabric that is rugged, absorbent, and warm.

How Many?

Brick Fortress

One of the largest brick structures in the United States is Fort Jefferson, which guards remote and tiny Garden Key, off the Florida coast. Sixteen million handmade red bricks went into the 15 m (50 ft.)-high, 2.4 m (8 ft.)-thick walls.

Leather

Long before ancient people learned how to weave fabrics, they wore animal skins to keep warm. Tanning is the process that turns an animal skin, or hide, into leather. When properly treated, leather can last for centuries. Today, leather is found in fine clothing, purses, saddles, and sports equipment like this soccer ball.

Then & NOW!

Concrete

Concrete has been used in construction for thousands of years. Bakery ovens in ancient Pompeii *(top left)* were built with an early form of concrete made from volcanic ash. Modern concrete, a mixture of sand, gravel, **cement**, and water, is reinforced with steel rods for even greater strength. The dazzling Guggenheim Museum in New York City *(bottom left)* consists of poured, reinforced concrete.

Ceramics and Glass

Ceramics are strong substances made by heating and shaping clay or other materials from the earth. There are two types of ceramics; one is molded first and then baked, the other is heated first and then poured or molded into its final form.

Early humans combined clay with water, formed the wet mixture into container shapes, and baked them in an open fire. This kind of pottery, which dates back to about 7000 BC, was handy for storing food and carrying water. Pottery today is heated in a special high-temperature furnace called a kiln.

Sometime before 2000 BC, people learned how to make glass. It has since become one of the most important types of ceramics. Although we usually think of glass as being clear, its main ingredient is silica sand! Today most glass is made in factories, but the finest pieces are still created by glassblowers, who shape the **molten** silica into beautiful designs.

Fiber Optics

Glass can be spun into long threads as thin as a cat's whisker. These transparent strands, called optical fibers, are very useful. Flexible fiberoptic cables can bend light around corners and into hard-to-reach places. Doctors use fiber optics to see inside a patient's body.

Nature's Glass

Did you know that glass can be found in nature? A shiny substance called obsidian is created when molten lava containing a high percentage of silica cools very quickly. Obsidian is usually black in color but can also be red, brown, green, or clear. In the photo at right, bands of ancient obsidian sparkle in California's Inyo National Forest. Prehistoric cultures carved lumps of obsidian into razor-sharp arrow- heads and spearpoints, as well as ceremonial figurines. Today, surgeons use obsidian scalpels for very delicate operations.

How Glass Is Made

Silica sand, pictured at top left piled outside a glass factory, is mixed with soda ash and limestone and heated until it becomes molten. The soft, hot glass is shaped and then cooled. Sheet glass for windows is rolled onto a conveyor belt *(center left)* in long sheets and later cut into smaller pieces.

Many other glass products, such as light bulbs and bottles, are made in the factory by pressing liquid glass into a mold. But fine glass ornaments are still shaped by hand, using the ancient art of glassblowing. To make a glass piece, the glassblower holds a metal blowpipe with a glob of molten glass on one end. He then turns the pipe and blows into it at the same time *(bottom left)* to create almost any shape.

Special Glass

Safety Glass

Ordinary glass shatters into sharp, jagged pieces. But when safety glass breaks, the blunt-edged fragments cling together *(right)*. Toughened glass, found in car windshields, is made under very high heat. Another type of safety glass, laminated glass, consists of a plastic layer between two sheets of glass.

Fiberglass

Fiberglass is made by forcing molten glass through tiny holes in the bottom of a furnace to create fine threads. These threads can be made into fabrics that are lightweight and strong. They protect against heat and fire as well, making fiberglass perfect for firefighters' clothing *(above)*.

Adding Metal to Glass

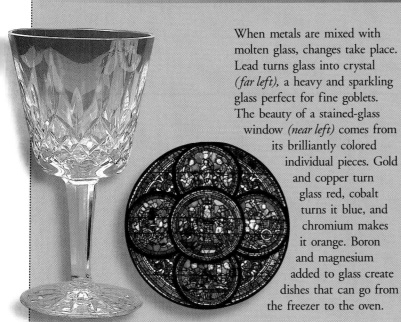

When metals are mixed with molten glass, changes take place. Lead turns glass into crystal *(far left)*, a heavy and sparkling glass perfect for fine goblets. The beauty of a stained-glass window *(near left)* comes from its brilliantly colored individual pieces. Gold and copper turn glass red, cobalt turns it blue, and chromium makes it orange. Boron and magnesium added to glass create dishes that can go from the freezer to the oven.

Ceramics: Baked Clay

The word "ceramics" comes from a Greek word meaning "earthenware." Some people still make plates, cups, and bowls in the traditional way, shaping wet clay on a spinning wheel *(bottom left)* and firing it in a small oven. But most dinnerware is produced in large potteries with kilns *(bottom right)* that can bake hundreds of pieces at a time. Today, clay ceramics are found in a wide variety of everyday items, from bricks and bathroom tiles to spark plugs and coffee mugs.

Metals Essential Elements

M etals are chemical **elements** found in Earth's crust. Of the roughly 80 known metals, very few occur in a pure state. They are combined instead with minerals and other elements in rocklike deposits called **ores.** The ore must be heated to a high temperature to separate and extract the metal in a process called **smelting.** The most common metals are aluminum—found in an ore called bauxite—and iron.

Metals are durable and are good **conductors** of heat and electricity. They can easily be shaped into many different forms, from long, thin wires to sturdy beams or flat sheets. Also, most metals will bend rather than break under pressure. All of our modern conveniences depend on metals: Automobiles, airplanes, skyscrapers, and household appliances could not exist without them.

What Are Precious Metals?

T he beauty of rare precious metals like gold *(top right)*, silver *(bottom right)*, and platinum makes them valuable. They are often found in jewelry, coins, and decorative items. But each of these metals also has more practical uses. Thin sheets of gold, which reflect light and heat, protect space-craft from the glare of the Sun. Some computer parts contain silver, an excellent conductor of electricity. Platinum can withstand high temperatures and is used in the manufacture of thermometers for furnaces.

From Old to New

What do an empty soda can, a worn-out washing machine, and a rusty old automobile have in common? They can all be **recycled!** Scrap metal from discarded items can be melted down and made into brand-new products. Not only is this cheaper than manufacturing new metal, it also reduces air-pollution and **landfill** problems.

It takes three times more energy to convert iron ore into steel than to recycle scrap steel like the compacted blocks shown above. This is why all steel made today contains some scrap steel. Many of the parts in a new car are actually made from recycled old cars!

Name?

Iron

Iron is never found pure in the Earth's crust, but a nearly pure form comes to Earth from space. Most meteorites are nearly 90 percent iron, which means that very little purifying is required. These chunks of space rock were used by the Inuit and other early cultures as a source of metal to make knives and other tools. In fact, the word for "iron" means "metal from the sky" in several early languages.

Steel, the Marvelous Metal

Steel is one of the most important metals in the modern world. It is strong, inexpensive to make, and easy to shape. In steel mills, workers take **molten** iron *(top left)* and put it into a **blast furnace.** Then carbon is added to the iron, and oxygen is blown through the mixture, creating the **alloy** known as steel.

Steel beams form the metal skeleton that supports the enormous weight of modern skyscrapers *(middle left).* But steel does have one drawback. When exposed to air and water, it rusts. Adding chromium to normal steel creates a nonrusting alloy called stainless steel, which is perfect for the blades of ice skates *(bottom left).*

Would You Believe?

Stronger Than Steel

Although a spider web looks very delicate, the individual threads are stronger than strands of steel of the same thickness. Organs called spinnerets in the spider's abdomen produce a liquid silk that hardens into an incredibly durable and flexible web.

What Was the Iron Age?

About 1000 BC a great advancement in human history took place in the Middle East and in parts of Europe. People learned to smelt iron by heating the ore to extremely high temperatures—more than 1,500°C (2,700°F)—to extract the pure metal. That period was later called the Iron Age, a time when the use of heavy, almost unbreakable metal tools and weapons became widespread.

What's Ore?

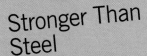

Any combination of metal and mineral contained in rock is called ore. In the photo at right, the deep red streaks of color in the earth indicate that iron ore is present. Aluminum is found in the mineral bauxite *(below),* which is also an ore. Ores are mined either on the surface from exposed deposits or underground. In open-pit mining, the ore is loosened by blasting and then removed, creating a terraced hole in the earth.

What Are Plastics?

Plastics are everywhere, in hundreds of items that we use every day, from toys and food containers to Rollerblades and computers. They are inexpensive, lightweight, waterproof, and easy to clean. They can be transparent or dyed to any color imaginable. Some are tough and rigid, like a bicycle helmet, whereas others are as springy and flexible as the bristles of a hairbrush.

Plastics are not found in nature and are therefore called **synthetic** materials. They are produced in a factory and made from chemicals that come from oil, natural gas, or coal. These materials from the earth are separated and processed in refineries or laboratories. Early plastics were invented as substitutes for natural materials that had become difficult or expensive to obtain, like tortoiseshell and ivory. During the 19th century, it became common to see buttons, knife handles, and piano keys made of plastic. Today it is almost impossible to imagine life without plastics. Look around you right now. How many different objects made of plastic can you see?

Types of Plastic

There are hundreds of different kinds of plastic, but they all fall into two types. Thermoplastics become soft and elastic when heated and harden again when cooled. Thermosets are permanently rigid and cannot be softened or remolded.

Products that are made from thermoplastics include beach-balls and sandwich bags. Because they will not melt when they're exposed to heat, thermosets are good for making products like hair dryers and computer disks.

Turning Oil into Plastic

Most plastics are made from petroleum. At oil refineries crude oil is heated to a high temperature, then sent to a refining tower, where it separates into various liquids, including gasoline and kerosene. Another liquid is processed further to make **molten** plastic. An extruder squeezes the plastic through holes, creating long, thin strands. A cutter slices the strands into chips that are later melted down and molded into different products.

Crude oil is heated to a very high temperature.

The oil separates into liquids at a refining tower.

One of the liquids is processed to make molten plastic.

The molten plastic is sent through an extruder to harden.

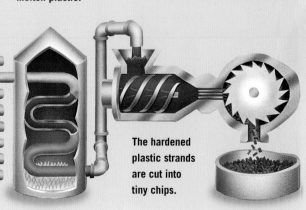

The hardened plastic strands are cut into tiny chips.

Early Plastics

PARKESINE

The first plastic, called Parkesine, was invented in 1855 by English chemist Alexander Parkes. He mixed vegetable oils and organic solvents with cellulose, which comes from the woody part of plants. The mixture resulted in a solid lump of material that would soften when heated. Although Parkesine won awards and was shown at the Great International Exhibition of 1862 in London, England, it was never widely used. One example is the plaque at right.

CELLULOID

The next plastic to emerge was a huge success with the public. John W. Hyatt, an American inventor, improved on Parkes's invention in 1869 with a plastic called celluloid. Celluloid first turned up in common items like combs, and soon found its way into the manufacture of toys, including the dolls at left. Later celluloid became the main ingredient

in movie and photographic film. Celluloid is still used today in small items such as table tennis balls.

Artificial Fiber

The first artificial fiber, called nylon, was developed in 1935 by American chemist Wallace H. Carothers. He was searching for a less expensive alternative to silk. Nylon is made by heating chips of plastic to 260°C (500°F), then squirting the hot, melted plastic through tiny holes, a process known as extrusion. The threads of nylon dry and set hard almost immediately. They are spun into a long yarn that is strong, flexible, and resistant to heat, oils, grease, and water. Nylon can be found in clothing, tents, carpets, and tires. At the 1939 World's Fair in New York, a knitting machine that produced nylon stockings (below) proved to be a popular exhibit.

BAKELITE

Bakelite was the first completely synthetic plastic and the first that could hold its shape when subjected to high heat. Belgian-born chemist Leo H. Baekeland named the substance after himself when he patented it in 1909. Bakelite became a popular material in telephones, pot handles, and car ignition systems. Although newer forms of plastics have replaced Bakelite for the most part, it is still widely used in automobiles and radios.

Plastics Close up

Plastics play an important role in modern technology. Just like the long-ago Iron Age, today's "plastics age" has transformed the world in many ways. Modern uses of plastics include spare parts for the human body and fibers that can stop a bullet, a chain-saw blade, or a white-hot flame.

Plastics have many advantages over other materials, but their strength and durability can also be a problem: Plastics are not biodegradable, which means they do not decompose or rot. This makes them hard to dispose of. Things that we use for a few minutes and then throw away could remain in the environment for centuries. In the United States alone, more than 14 million tons of plastic are thrown away every year. Turning **recycled** plastics into products such as clothing and sports equipment is one solution. Developing biodegradable plastics is another.

Plastics from Plants

Scientists have developed a plant that grows plastic. They inserted a special type of bacteria into a small weed called the mouse-eared cress *(right),* and the plant grew tiny grains of a biodegradable plastic. This plastic, known as PHB, is similar to the plastic in milk containers. Plants that produce plastic may help us reduce our dependence upon Earth's oil supply. Maybe someday all our plastics will come from the garden!

Recycling Plastic

Where do all those wonderful plastic items go when we are finished with them? For a long time they went straight from our trash cans into **landfills**, where it may take hundreds of years for a plastic bottle to disintegrate. These days, recycling can turn discarded plastics into many things, from artificial fleece vests *(top left)* to the felt of tennis balls *(far left)*. At left, a mountain of plastic bottles sits outside a recycling facility.

Plastics in the Body

For centuries, people have created replacements for body parts that were damaged, diseased, or missing. Today many of these artificial parts are made from tough plastics. False teeth, called dentures, are made of the hardest plastics of all. A modern plastic eye can look

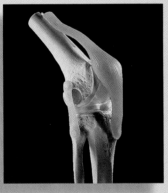

very real, although the wearer cannot see through it. Plastics used deep inside the body are called implants, like the artificial knee joint at left.

This natural elastic is made from the sap of rubber trees that grow in tropical countries. The sap is a milky liquid called latex. Workers make several thin cuts in the bark of the tree, and the latex oozes out *(top right)*. The latex is collected in a container, then sent to a factory for processing. There the liquid is hardened and made into a wide variety of products, from balls and rubber bands to tires and raincoats.

Rubber is waterproof and can be stretched to several times its own length without breaking. The factory at bottom right is making rubber laboratory gloves. The ceramic glove forms are dipped in a bath of latex, then put into an oven so the rubber will dry.

Lifesaving Plastics

Did you know that plastics can save lives? American research chemist Stephanie Kwolek *(left)* created a fiber called Kevlar that is five times stronger than steel but very lightweight. This miracle fiber is flameproof, bulletproof, and cut resistant. Loggers wear Kevlar gloves to protect their fingers. Kevlar is also used in soldiers' helmets, radial tires, space vehicles, and many other items.

Like Kevlar, Spectra is another ultrastrong and lightweight fiber that is perfectly suited for the bulletproof vests used by police officers and even police dogs *(right)*. Materials like Kevlar and Spectra have saved hundreds of lives and prevented countless injuries from cuts, burns, and bullets.

New Materials

A whole range of advanced, man-made materials have transformed modern life. Today's high-tech ceramics, plastics, carbon fibers, and composites create products that are stronger, safer, and more efficient than ever before. New wear-resistant ceramics can be found in everything from artificial bone implants to scissors that never need sharpening.

Very thin fibers of pure carbon have proved to be one of the most practical materials ever invented. Automobile and bicycle parts made of carbon fiber are stronger, lighter, and less expensive than traditional metal parts. Sporting equipment like golf clubs, fishing rods, and tennis rackets is made of carbon fiber **composites.**

Composites are created when extremely stiff particles or fibers are combined with tough materials like plastics, metals, or ceramics. Many advanced materials, including composites and memory metals, were originally developed for use in the space program but have found their way into everyday use.

Weightless Wonder

Carbon fibers are only about one-seventh the width of a human hair but are three times as strong as steel. They can withstand pressure of 70,000 kg/sq. cm (1,000,000 lb./sq. in.) yet are extremely lightweight. The Ultimate Bike *(below)* has a carbon fiber frame and weighs a mere 5 kg (11 lb.). In 1995 this bike set the world speed record for bicycles at almost 333 km/h (208 mph). The tires are a combination of silk and Kevlar fiber, and they're filled with helium instead of air.

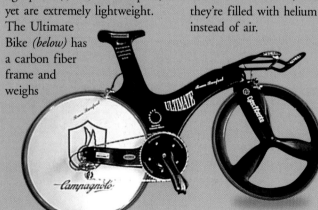

A n artificial device to replace a missing part of the body is called a prosthesis. The first prosthetic devices were fashioned from wood, metal, or ivory. Today they are made of the strongest and most modern **synthetic** materials. The man at right lost his legs in a truck accident. He now has prosthetic legs made from titanium **alloys,** carbon fibers, and a type of plastic called polyethylene. Through many hours of physical therapy, he relearned the art of walking and running.

Space Technology Spin-Offs

The space program produced many inventions that have become a familiar part of our everyday lives. One of these spin-offs is shape-memory alloy, also called memory metal. These metals can be bent out of shape, but when heated will "remember" their original shape. Memory metals were originally developed for a space antenna that was shipped into orbit as a tiny package, then opened up to full size by the warmth of the Sun. Flexible, lightweight eyeglass frames are one down-to-earth benefit of memory metals.

Advanced Ceramics

Adding certain chemical compounds to ceramics makes them much stronger, harder, and more heat resistant than traditional bricks and pottery. Some advanced ceramics outperform iron and steel in tests of stiffness and strength. Ceramic scissors *(right)* stay razor sharp and never rust. Body fluids do not dissolve these superhard materials, so they make ideal artificial bone and joint implants. Ceramic tiles that can withstand temperatures of 1,260°C (2,300°F) cover the underside of the space shuttle *(below)* and prevent the ship from burning up when it reenters Earth's atmosphere.

Micromachines

Scientists use advanced materials like silicon to build machines that are so small they can't be seen with the naked eye. In the photo below, magnified more than 100 times, a tiny spider mite dwarfs a gear mechanism the size of a grain of pollen. In the photo at bottom, magnified 25 times, the gear is barely visible on the ant's head. Micromachines are currently used to trigger automobile air bags, transmit the ink in ink-jet printers, and sense ice on the wings of airplanes.

As small as micromachines are, imagine something that's even smaller, one-billionth the size! Some scientists believe that one day we will be able to build devices called nanomachines. In science, *nano* means "one-billionth." Smaller than a human cell, these nanomachines could be programmed as tiny medical robots. They would be sent into a patient's bloodstream to kill cancers and viruses and clear blocked arteries.

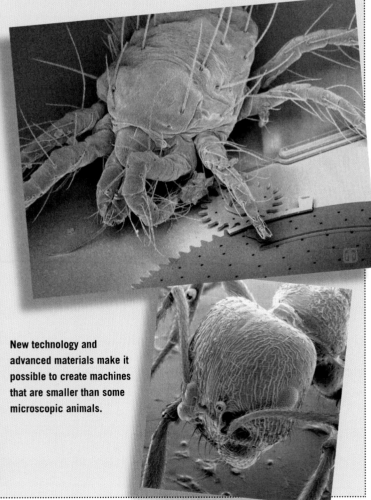

New technology and advanced materials make it possible to create machines that are smaller than some microscopic animals.

Tall Structures

Until the late 19th century, large buildings were made either of brick or of stone. Brick buildings will stand for centuries, and stone structures, like the medieval cathedrals of Europe, can last even longer. The one big problem with brick and stone is their extreme heaviness. Buildings only a few stories high have to have massive walls on the lower floors to support the weight of the upper floors.

In the 1880s architects devised a rigid metal framework that was lightweight but sturdy enough to hold up the roof, walls, and floors of a building. With this "skeleton" of iron and steel, walls could be made much thinner and structures could be built much taller than ever before. We call these really tall buildings "skyscrapers." The very first of these, the 10-story-high Home Insurance Building, dominated the skyline of Chicago, Illinois, when it was built in 1885.

Let's Compare

Skyscrapers

For 3,800 years, the tallest structure in the world was the Great Pyramid of Khufu in Egypt. It was 147 m (482 ft.) tall. In the Middle Ages, the spires of Gothic cathedrals reached even higher. Since the late 19th century, a succession of sky-scrapers have held the title. The illustration below shows some of the tallest buildings in the world today. Three of the six were built in the 1990s, so the list will probably keep growing and changing.

1. **Petronas Twin Towers** (452 m; 1,483 ft.)
2. **Sears Tower** (443 m; 1,454 ft.)
3. **Jin Mao Building** (421 m; 1,380 ft.)
4. **World Trade Center** (417 m; 1,368 ft.)
5. **Empire State Building** (381 m; 1,250 ft.)
6. **Central Plaza** (374 m; 1,227 ft.)

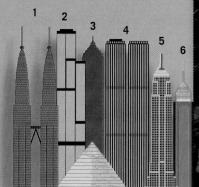

Twin Titans

The tallest building in the world is actually a pair of identical structures. The 88-story Petronas Twin Towers in Kuala Lumpur, Malaysia, are gleaming cylindrical giants linked in the middle by a glass bridge. The towers contain enough steel to build more than 35,000 cars! The buildings may not hold the world record for much longer —several taller buildings are planned for the near future.

What Made Them Possible

Early elevators were very dangerous; if a cable broke, the car would plunge to the bottom of the elevator shaft. American engineer Elisha G. Otis invented a new type of elevator *(right)* that made skyscrapers possible. In 1854, at a public demonstration of his new invention, he stood in the car while someone cut the cable. An automatic safety device locked into place and kept Otis and the elevator from falling.

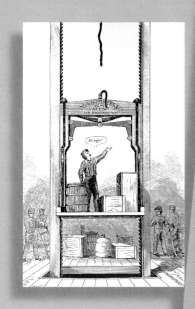

A new breed of construction worker was required to build the towering, open framework of a skyscraper *(left)*. Welding joints or hammering rivets while perched on a 20 cm (8 in.)-wide beam 40 stories above the ground *(bottom)* was not for everyone. Ever since the earliest skyscrapers went up, many of these "high-iron workers" have been Native Americans.

Skyscraper

The Home Insurance Building, designed by American architect William L. Jenney, was the first true skyscraper. Built in Chicago, Illinois, in 1885, it was 10 stories high and had a safety elevator. Most earlier buildings had six stories or fewer. The Home Insurance Building was the first building that used a framework of iron and steel to help support its weight. Jenney's innovative steel framework design paved the way for future buildings to soar to incredible heights. Some skyscrapers today are more than 100 stories tall!

Towers of Trash

On a small plot of land in Los Angeles, California, stands one man's monument to trash. Italian immigrant Simon Rodia started his project in 1921. He spent the next 33 years building a group of towers *(top right)*, one of which grew to more than 30 m (100 ft.) tall. Bottle caps, broken dishes, seashells, and other discarded materials adorn the towers' concrete-coated steel rods *(bottom right)*. Many of the walls are imprinted with items Rodia found in junkyards, like automobile gears, straw baskets, and faucet handles. Called the Watts Towers, these amazing structures may be the largest work of art ever created by one person. The towers have been designated a historic monument by the U.S. Congress.

Flying Buttresses

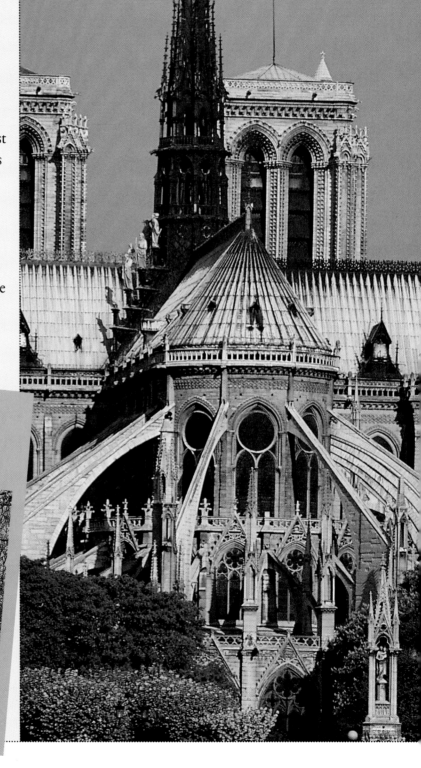

What keeps tall buildings from toppling over? Engineers use some common principles of physics together with certain structural elements to keep them standing. Exterior supports like flying buttresses and firm foundations combined with physical **forces** such as tension, **compression,** and shear in the interior have been used for centuries to steady buildings.

When all these elements are properly matched, the result is a strong, magnificent structure, like a Gothic cathedral or the Eiffel Tower, that can withstand the test of time. Architects and engineers ignore these principles at their peril, as is clearly illustrated in a famous but flawed medieval Italian tower.

A flying buttress is an arch that presses against the outside of a heavy stone building, preventing the walls from buckling outward. Developed in the 12th century, these delicate-looking yet strong arches were used on the Cathedral of Notre Dame in Paris, France *(right).*

Strange But TRUE!

Termite Towers

Can you imagine tiny termites building towers up to 6 m (20 ft.) high? If termites were as tall as humans, their towers would be four times the height of the Empire State Building! Made of mud and saliva, these insect skyscrapers are as hard as concrete and can house millions of termites. Although the average termite's life span is only about a year, a tower might survive for up to a century.

Stressed Out

The three main types of force, or stress, that affect tall structures are tension, compression, and shear. The red arrows in the illustrations here show the direction of the force in each case.

Tension

When a force pulls a material, it is under tension and tends to stretch. At right a weight pulls a steel cable taut, making it rigid.

Shear

Shear, or twisting, is called the sliding force. The steel bolt above holds together a column and a girder, each pulling in an opposite direction.

Compression

Compression is the stress that pushes together and shortens material. A weight placed on a column will drive it downward against the ground.

A Firm Foundation

Like a tree needs its roots, tall buildings need solid foundations to support their massive weight and resist the strong winds that might topple them. Columns of reinforced concrete that support a skyscraper's weight, called piles *(bottom left)*, extend through the soil to a layer of solid rock.

The Leaning Tower of Pisa's foundation did not reach all the way to bedrock. Soon after construction began, in 1174, the tower started tilting in the soft, unstable soil. Still, it has stood for more than 800 years, and currently leans about 5 m (17 ft.) from the vertical *(top left)*.

Triangle Power

The triangle is an important shape in both natural and man-made structures. It is very strong because no one side can bend away from the other two. The only way to change a triangle's shape is to break it. Large structures like bridges are often made up of many triangles joined together to form a truss. When a giraffe bends over to drink, it spreads its front legs to form a stable triangle with the surface of the ground *(near right)*. The Eiffel Tower in Paris, France, has hundreds of triangles in its design, making it a very stable structure *(far right)*.

Bridges Dams and Tunnels

Bridges carry railroads and highways safely over deep gorges, bodies of water, and even each other. Dams collect and store water, preventing floods and generating electricity. Tunnels take us through mountains, under cities, and beneath riverbeds. Other kinds of tunnels transport water, gas, and electricity beneath our feet and remove waste materials from our houses.

All three of these engineering structures were first built by ancient cultures using natural materials. The earliest bridges were probably flat stones or tree trunks laid across a stream. People also wove fibers together to make rope bridges. Irrigation dams made of earth or stone were built 5,000 years ago across the Nile River in Egypt and the Tigris River in Mesopotamia. The civilizations of India, Greece, and Rome dug tunnels to bring drinking water to cities. On the island of Sámos in Greece, a tunnel dug into a mountain more than 2,000 years ago is still visible.

Today, bridges, dams, and tunnels are built all over the world using modern construction techniques and materials. As with buildings, new records are regularly being set for the tallest, biggest, and longest. Suspension bridges stretch across ever greater distances, some concrete dams approach the height of the world's tallest skyscrapers, and an amazing tunnel has connected two countries—England and France—beneath a 50 km (31 mi.) stretch of open sea.

How **Long?**

Akashi-Kaikyo Bridge

The longest suspension bridge in the world today is the Akashi-Kaikyo Bridge in Japan. The 3.9 km (2.4 mi.)-long bridge was built to withstand earthquakes and strong winds. In fact, it was nearing completion when a 1995 earthquake shattered the nearby city of Kobe but left the bridge untouched.

Types of Bridges

The most common bridge type is the simple beam bridge. An arch bridge can span a wider gap than a beam bridge and is much stronger. Each section of a cantilever bridge has its own support in the river. The sections are joined by a span. A bascule bridge is hinged so that sections can be raised for ships to pass. Of all bridge types, the graceful suspension bridge spans the greatest distance between its towers.

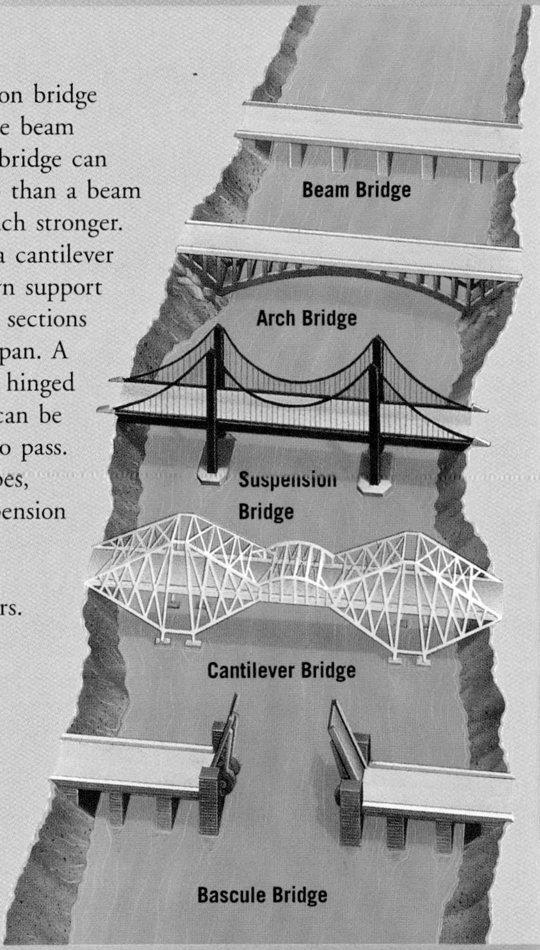

Beam Bridge

Arch Bridge

Suspension Bridge

Cantilever Bridge

Bascule Bridge

Types of Dams

Dams can be small mounds of soil or huge concrete barriers. Since the beginning of recorded history, people have built dams across rivers and other bodies of water to store or divert water. Modern dams can release water in powerful streams that drive turbines to produce electricity. This is called hydroelectric power. Illustrated below are the four basic types of dams.

Buttress Dam   **Embankment Dam**   **Concrete Gravity Dam**   **Arch Dam**

The Channel Tunnel

For more than a century, England and France had discussed building a tunnel under the English Channel. Work began in 1880, but the project was soon stopped. The same thing happened in the early 1970s. Finally, another tunnel was started in 1987 and completed

in 1994. It is 50 km (31 mi.) long, of which 38 km (24 mi.) are underwater. The rotating head of the huge boring machine *(left)* chewed through the bedrock like a giant drill. The Channel Tunnel actually consists of three separate tunnels, two for trains and the third for maintenance and emergencies.

Shipworm

In 1825, undersea tunneling became much safer when British engineer Marc I. Brunel developed a new kind of boring machine. He got his inspiration for this device by observing the shipworm *(right)*, which is actually a mollusk that loves to bore into wood. The shipworm rotates the sharp edges of its shell, producing tiny shavings of wood powder. It then eats the powder and excretes a substance that forms a hard lining in the tunnel. Like the shipworm, Brunel's boring

machine moved the rock and soil back through the tunnel and disposed of it.

In the Path of Nature

Tacoma Narrows Bridge

Earthquakes are frightening natural events that can do terrible damage. Every year, about 20 serious earthquakes occur around the world. Just a few seconds of violent shaking destroys buildings and causes millions of dollars in damage—and sometimes many casualties.

As skyscrapers grow taller and bridges longer, the force of strong winds also becomes a concern. Every structure has what is called a resonant frequency *(page 75)*. If a structure vibrates at its resonant frequency—in a high wind, for example—it may tear itself to pieces.

In the past, builders simply tried to make tall structures rigid enough to withstand earthquakes. But without any flexibility, they could literally be shaken apart. In earthquake-prone areas, engineers now include slightly flexible materials in their designs so that the structures can sway and bend a little without collapsing. These buildings bend like trees in the wind and ride out earthquakes like a cowboy on a bucking horse.

Taiwan Earthquake

Taiwan's most powerful earthquake in more than a century hit the small island nation on September 21, 1999. It left more than 2,200 people dead, 6,500 injured, and 100,000 homeless. The quake knocked over almost 6,000 buildings. Some taller buildings ended up leaning against their smaller neighbors like dominoes.

One apartment complex collapsed straight down on its foundation like an accordion, squeezing 12 floors into just four. The 12-story building at left simply toppled over onto its side.

In July 1940, a new suspension bridge opened over Tacoma Narrows in Washington State. Four months later, when a strong wind blew up, the bridge began to twist wildly *(1)*. The span moved up and down, trapping a lone car on the bridge *(2)*. The driver managed to escape by jumping out of the car and crawling away. The roadbed finally shattered under the stress, sending hundreds of tons of steel and concrete into the water below *(3)*. Bridge engineers had failed to take into account the powerful, gusty winds that often blow through Tacoma Narrows and that caused the bridge to vibrate at its resonant frequency.

Then & NOW!

Resisting Earthquakes

An earthquake can turn a rigidly constructed building into rubble within seconds. American architect Frank Lloyd Wright understood this when he redesigned the Imperial Hotel in Tokyo, Japan *(top)*. In 1923 the hotel survived one of Tokyo's worst earthquakes ever. Even though 5,000 buildings were

shattered, the Imperial Hotel suffered not so much as a broken window! The hotel's foundation rested on a shallow layer of firm soil atop a shaky layer of mud, allowing the building to move with the motion of the earthquake.

Buildings constructed today in San Francisco, California,

must be earthquake resistant, like the pyramid-shaped Transamerica Building at right. It is designed to sway within planned, safe limits during a quake. And because the 260 m (853 ft.)-high building is much wider at the base than at the peak, it will not topple over.

Against the Wind

Because wind pressure increases with altitude, winds blow much harder against the upper floors of a skyscraper than at ground level. This creates enormous pressure on the base. Engineers cope with these stresses in a variety of ways. A central core (1) can act as a sturdy backbone to stiffen the building. A suspended building (2) hangs by cables from a central core and is built to sway a little. An

externally braced building (3) has diagonal supports on its outer walls. A tuned mass damper (4) is a huge concrete block that rides an oil-slickened pad on the roof of the building. It is computer controlled and can slide in any direction to counterbalance the force of the wind.

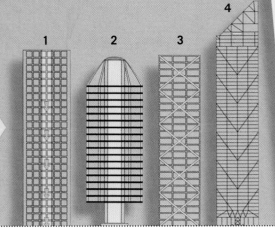

1   2   3   4

Onion Domes

Many churches in Russia and eastern Europe have domes shaped like an onion. They stand atop tall cylindrical structures called drums. These domes are not just ornamental; they are also functional. The onion shape helps them shed the heavy snow and rain that fall in these northern countries. The brightly patterned onion domes at right crown St. Basil's Cathedral, located in the heart of Moscow on Red Square. The cathedral was constructed in the 16th century. Domes made from materials as diverse as mud,

stone, concrete, and plastic have been used in architecture for thousands of years.

ENGINEERING   115

Classical Physics

Since the beginning of time, people have wondered about the natural world. Over the millennia, scientists slowly added to the understanding of the relationships between matter and **energy** in the universe, seeking to arrive at universal laws that describe the workings of the physical world.

Until 1900 scientists increased the knowledge about what is now known as "classical physics," which covers **gravity** and motion, heat, sound, light, magnetism, and electricity. Each of the scientists included in this timeline contributed significantly to our understanding of the physical world, laying the groundwork for the mind-boggling ideas and amazing discoveries that would come in the 20th century.

In ancient times, people thought the Sun, the Moon, and the stars lay underneath a crystal dome that covered the flat Earth.

Launching with Levers

Among the many scientists who studied at the famous library of Alexandria in Egypt was Archimedes *(right)*. He became famous for his work in geometry (with circle, sphere, and cylinder) and explained the principles of **mechanics** with simple machines. He once used a single lever to launch a warship for the king of Syracuse.

Newton, the Know-It-All

Before he was 25 years old, British physicist Isaac Newton had invented the mathematical method called calculus, outlined a theory of light and color, and discovered how gravity works. Newton also described the three laws of motion and invented the reflecting telescope. One of the most brilliant scientists who ever lived, Newton was knighted for reforming the English currency and coinage systems.

300 BC—AD 1600

c. 307 BC- AD 640 The library of Alexandria in Egypt draws astronomers, philosophers, mathematicians, and other scientists from all over the ancient world.
AD 1543 Copernicus proposes that the Sun is the center of the solar system.

1601—1699

1609 Galileo confirms that the Sun is the center of the solar system. He experiments with inclined planes and the **velocity** of falling objects.
1687 Isaac Newton publishes *Principia,* describing universal gravitation and the laws of motion.

1700—1755

1705 Francis Hauksbee proves that sound will not travel in a vacuum.
1738 Daniel Bernoulli studies the behavior of gases.
1752 Benjamin Franklin shows that lightning is a form of electricity.

Timeline Events

A Charge from Electricity

Nikola Tesla revolutionized the study of electricity through his experiments with alternating **current.** One is shown here in a composite photo. Alternating current allows electric power to travel great distances from its source. Tesla applied for more than 700 patents in his lifetime, including one for a transformer

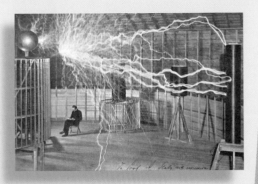

called the Tesla coil that produces high-frequency, high-voltage electricity and is still used in some common appliances.

Finding Out about Electrons

"Like raisins in a cake" is how British physicist Joseph J. Thomson described the placement of **electrons** in **atoms.** He discovered the electron while observing what happened to a gas when it became a **conductor** of electricity. The existence of electrons had been theorized several decades earlier, but Thomson was the first to observe their behavior. Understanding electrons has helped to power computers, lasers, x-ray machines, and television tubes during the 20th century.

The Nature of Light

Experiments with light and color led Scottish scientist James Clerk Maxwell to discover that light travels in **oscillating** waves of electric and magnetic energy, and that visible light is only one type of electromagnetic radiation.

In 1860 Maxwell's experiments with color photography produced the first color photograph, which showed a scrap of tartan like the one at left.

A Scientific Pioneer

Winner of two Nobel Prizes, Marie Curie (1867-1934) was an early investigator of radioactivity and discovered the radioactive elements polonium and radium with her husband, Pierre. At the outbreak of World War I in 1914, she helped outfit ambulances with x-ray equipment and rode to the front lines to treat patients.

1756—1845

1798 Benjamin Thompson (Count Rumford) finds a direct relationship between heat and mechanical energy.
1800 Alessandro Volta invents the electric battery.
1831 Michael Faraday discovers the production of electric current through magnetism.

1846—1869

1851 William Thomson (Lord Kelvin) describes the first two laws of thermodynamics.
1853 William Rankine introduces the concept of potential (stored) energy.
1869 Dmitry Mendeleyev creates the periodic table of **elements.**

1870—1890

1873 James Clerk Maxwell discovers that light is a form of electromagnetic radiation.
1885 Carl Benz and Gottlieb Daimler develop an internal-combustion engine.
1886 Heinrich Hertz generates and detects radio waves.

1891—1899

1895 Wilhelm Roentgen discovers x-rays.
1896 Marie and Pierre Curie begin experimenting with **radiation.**
1897 J. J. Thomson discovers electrons in an atom.
1898 John Townsend measures the charge of an electron.

Modern Physics

In the 20th century, physicists sought to explain the universe on the very small and the very large scale. In 1900 Max Planck formulated the quantum theory, suggesting that **radiation** was made up of "packets," or quanta, of **energy,** which was emitted in leaps rather than as a continuous emission. Albert Einstein's special and general theories of relativity described how space and time, matter and energy, and gravitation behaved on a universal scale. These ideas triggered important new research.

Double-Edged Atom

Learning about the atom—such as the cluster of seven uranium atoms in the photo at left—has had a profound impact on the 20th century. With the discovery of **subatomic particles,** physicists have been able to delve deeper into the mysteries of the universe, but they also unleashed the technology for building **nuclear** weapons, capable of almost inconceivable destruction.

Richard Feynman

Richard P. Feynman had an extraordinary gift for visualizing complex problems. He shared a Nobel Prize for developing the theory of quantum electrodynamics, which used the quantum theory to explain the interaction between **electrons** and **protons.** Feynman was a popular public personality who helped make theoretical physics accessible to laypeople.

Theory of Relativity

In 1905, 26-year-old Albert Einstein revolutionized physics by suggesting that space and time are not separate, but instead are part of a four-dimensional continuum called space-time. Einstein's theory made great strides toward explaining how gravitation affects the universe. His famous equation, $E=mc^2$, proposed that energy and matter were essentially the same thing.

Timeline Events

1900—1914

1900 Max Planck formulates the quantum theory of matter.
1901 Guglielmo Marconi sends radio waves across the ocean.
1905 Albert Einstein formulates the special theory of relativity.
1909 Bakelite—the first completely **synthetic** plastic—is invented.

1915—1930

1915 Albert Einstein posits his general theory of relativity.
1916 Niels Bohr proposes a model of the **atom** based on quantum theory.
1926 Television is invented.
1929 Edwin Hubble proves the universe is expanding.

1931—1945

1935-1938 Radar is developed.
1938 Enrico Fermi builds the first nuclear reactor.
1939 The first FM radio station is built.
1942 The electron microscope is invented.
1945 First atomic bomb explodes.

A Theory of Everything

Confined to a wheelchair because of a neurological disease, Stephen Hawking is among many physicists working to synthesize ideas about **gravity**, electromagnetic forces, and nuclear and weak interactions into a theory of everything. Such a theory would explain the workings of everything in the universe, from stars to subatomic particles.

Hubble's Expanding Universe

Gazing at the sky from California's Mt. Wilson Observatory, astronomer Edwin Hubble discovered that contrary to what most astronomers thought, the Milky Way was just one galaxy among many millions in the universe. He further shocked the world by showing that the universe is expanding like a giant balloon and has been doing so for billions of years. Hubble's discovery enabled scientists to calculate the approximate age of the universe at 13 billion to 15 billion years. NASA's Hubble Space Telescope is named in his honor.

Atomic Power

Enrico Fermi *(bottom right)*, an Italian-born physicist, brought about the first controlled nuclear chain reaction under the stands of the football field at the University of Chicago *(center)* in 1942. During World War II, Fermi was among many brilliant scientists to work on the top-secret Manhattan Project, which culminated in the explosion of the first atomic bomb in 1945. Since then, nuclear weapons have played a major role in global relations.

Transistor Revolution

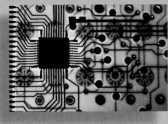

Transistors *(below)*, invented in 1948, can amplify **current** and switch it on and off. Millions of transistors packed onto microchips *(top left)* power today's computers, which have become an integral tool for work and leisure, as seen in the cybercafé at left.

1946—1959	1960—1975	1976—1985	1986—2000
1948 The transistor is invented. **1951** The first commercial computer is introduced. **1957** Sputnik, the first artificial satellite to orbit Earth, is launched.	**1960** The laser is invented. **1967** Home microwave ovens are produced. **1967** Jocelyn Bell discovers pulsars. **1969** Man lands on the Moon. **1970** Fiber optics is developed. **1971** The silicon chip is invented.	**1977** Apple II, the first personal computer, is introduced.	**1986** The Global Positioning System is introduced for commercial use. **1989** The World Wide Web is launched. **1990** The Hubble Space Telescope becomes operational.

Picture Credits

121

Glossary of Terms

Acceleration The rate of change in speed or direction of a moving object.

Acid A substance that releases hydrogen ions when it is dissolved in water. Acids have a sour taste and are neutralized by bases.

Acoustics Relating to sound, the science of sound, or hearing.

Alloy A material made of two metals or of a metal and a nonmetal.

Atom The smallest unit of an element that still retains the chemical properties of the element.

Atomic number The number of protons in an atom's nucleus.

Base A substance that releases oxygen and hydrogen when dissolved in water. Bases have a slippery, bitter taste and are neutralized by acids.

Blast furnace A large, intensely hot oven that is used to smelt metal ores.

Bond A connection between two atoms.

Cement A powder made of clay and other materials that are heated and then finely ground and used to make mortar or concrete.

Charge The amount of electric energy an object contains. An object with a negative charge has more electrons than protons, and an object with a positive charge has more protons.

Chemical An element or a compound produced in a process or used in a reaction.

Chemistry The science dealing with the structure and properties of matter and the reactions of elements and compounds.

Composite Made up of different parts.

Compound A substance made of two or more elements.

Compression The act of squeezing or pressing together; the opposite of rarefaction.

Concentration Strength.

Conductor Material through which heat or electricity can flow.

Converge To come together.

Current A flow of electric charge.

Density The ratio of an object's mass to its volume; the amount of material in a measurable unit.

Diverge To spread out from a common point.

Electric charge An amount of electric energy or an excess of electrons.

Electromagnetic waves Waves of electric and magnetic energy that travel through space at 297,600 km (186,000 mi.) per second, including radio waves, infrared, visible light, ultraviolet, x-rays, and gamma rays.

Electron A negatively charged particle that orbits the atom's nucleus.

Element Any substance composed of a single type of atom.

Energy The capability of doing work; the ability to make things move or change.

Enzyme A substance that aids chemical reactions.

Fiber Thin, threadlike materials, natural or man-made, usually capable of being spun into yarn.

Field The space or region in which a given force, such as magnetism or electricity, is active.

Focal length The distance between a lens or mirror of a telescope and the point at which it focuses.

Force The cause of motion or a change in the speed or direction of motion.

Foundation The base on which a structure is built.

Friction A force that resists motion between objects.

Fuel A material that is burned to produce heat or power.

Gravity The force responsible for the attraction of separate objects. The strength of the force depends on the mass and the distance between the two objects.

Interference The reinforcement or counteraction of two waves when they meet.

Ion An atom with a charge. Positive ions have lost an electron, and negative ions have gained an electron.

Landfill A waste-disposal system in which an area of land is used to bury garbage and trash under layers of earth.

Magnitude Size, quantity, or loudness.

Malleability The ability of a material to be hammered or pressed into a new shape without breaking.

Mass A measure of the total amount of material in an object.

Mechanics The science that deals with energy and forces and their effect on bodies.

Medium A substance, such as air or water, through which something is transmitted.

Molecule The smallest possible quantity of any substance that is made up of more than one atom; a particle containing one or more atoms.

Molten Melted.

Natural resources Anything that can be used by people and is provided by the natural environment, such as trees, minerals, or petroleum.

Neutral Neither an acid nor a base.

Neutron A noncharged particle in the nucleus of an atom.

Nuclear Related to the nucleus of an atom or energy from the nucleus of an atom.

Nucleus The center of an atom around which electrons orbit.

Orbit The path of an object revolving around another object; the path of an electron around the nucleus of an atom.

Ore A rock or mineral in which a metal is bound chemically to other metals or nonmetals.

Oscillate To swing back and forth in a set pattern.

Phase change A change in matter in which atoms are arranged differently because of a change in temperature or pressure; a change from solid to liquid or to gas.

Photon A packet of energy associated with a specific electromagnetic wavelength.

Pollution Material that harms the environment, making it less suitable for living organisms.

Pressure The force of air or water pressing against a surface.

Prism A clear, solid object with two nonparallel sides used to bend light or break it into its separate colors.

Propagate To transmit through a medium, such as air, water, or space.

Proton A positively charged particle in the nucleus of an atom.

Radiation Waves or particles of energy.

Rarefaction The regions in a compressional wave at which the particles are farther apart than in their resting state; the opposite of compression.

Reaction A chemical change.

Recycle To reuse or remove useful materials from used goods.

Reflect To bend or turn light waves or sound waves back from a surface.

Refract To bend or turn light waves or sound waves when they pass from one medium to another.

Resistance The opposition to the flow of electrons; an opposing force.

Resonance The frequency at which an object naturally vibrates.

Smelting The process in which a metal ore is heated with oxygen or another gas to isolate the pure metal.

Solvent A substance that can dissolve a compound.

Speed The rate of motion.

Static electricity A buildup on an object of either negative charges from the gain of electrons or positive charges from the loss of electrons.

Subatomic particles Particles that combine to make up atoms.

Suspension A substance in which particles are mixed but not dissolved.

Synthetic Man-made.

Toxic Poisonous or unhealthful.

Transformer A device that changes the voltage of an electric current.

Velocity The speed of an object in a specific direction.

Vibration Rapid and rhythmic movement back and forth.

Voltage The force exerted in accelerating electrons along a circuit to form a current. Voltage is measured in volts.

Volume The amount of space taken up by a substance.

Wavelength The distance from crest to crest or trough to trough of an electromagnetic wave or other wave.

Work The result of a force acting on matter and producing motion.

Index

Index

TIME LIFE BOOKS

Time-Life Education, Inc. is a division of Time Life Inc.

TIME LIFE INC.
CHAIRMAN AND CHIEF EXECUTIVE OFFICER: Jim Nelson
PRESIDENT AND CHIEF OPERATING OFFICER: Steven Janas
SENIOR EXECUTIVE VICE PRESIDENT AND CHIEF OPERATIONS
OFFICER: Mary Davis Holt
SENIOR VICE PRESIDENT AND CHIEF FINANCIAL OFFICER:
Christopher Hearing

TIME-LIFE BOOKS
PRESIDENT: Joseph A. Kuna
VICE PRESIDENT, NEW MARKETS: Bridget Boel
GROUP DIRECTOR, HOME AND HEARTH MARKETS:
Nicholas M. DiMarco
VICE PRESIDENT AND PUBLISHER, TIME-LIFE TRADE:
Neil S. Levin

Time-Life Student Library
PHYSICAL SCIENCE

EDITORS: Jean Crawford, Karin Kinney

Associate Editors/Research and Writing: Lisa Krause, Mary Saxton
Picture Coordinators: Daryl Beard, Julianne Moore, Anne Whittle

Designed by: Alicia Freile and Phil Unetic, 3r1 Group

Special Contributors: Susan Blair, Patricia Daniels, Mark Galan, Jocelyn Lindsay,
Jim Lynch, Terrell Smith, Elizabeth Thompson; Barbara Klein (index)
Senior Copyeditors: Judith Klein, Mary Beth Oelkers-Keegan
Correspondents: Maria Vincenza Aloisi (Paris), Christine Hinze (London),
Christina Lieberman (New York)

Consultants:
Richard E. Berg, Ph.D., has taught at the University of Maryland for over 33 years
and is presently the director of the Physics Lecture-Demonstration Facility. He
is coauthor with David G. Stork of *The Physics of Sound,* published by Prentice-
Hall, now in its second edition. For the past 18 years he has presented an exten-
sive series of physics demonstration programs called Physics is Phun, attracting
almost 5,000 visitors yearly to the university. Teaching interests have included
sound and light for introductory students and an honors course called Nuclear
Physics and Society, which addresses contemporary issues involving nuclear
physics. He has also written for the *Encyclopedia Britannica.*

Ed Matthews has taught eighth-grade physical science for 21 years. During this time,
he has completed multimedia projects for the Discovery Channel, developed probe-
ware labs for the Smithsonian Institution, and produced two television programs on
educational science topics for a cable television channel. He has given presentations
at both local and state conferences. He currently serves as a technology training
specialist for Fairfax County Public Schools in Fairfax County, Virginia.

School and library distribution by Time-Life Education, P.O. Box 85026,
Richmond, Virginia 23285-5026.
Telephone: 1-800-449-2010
Internet: www.timelifeedu.com

TIME-LIFE is a trademark of Time Warner Inc. and affiliated companies.

Library of Congress Cataloging-in-Publication Data
Physical science.
    p. cm. — (Time-Life student library)
  Includes index.
  Summary: Reviews fundamental scientific principles in areas such as simple machines,
electricity, magnets, lights, matter, chemistry, motion, and engineering.
  ISBN 0-7835-1359-3
  1. Physical sciences—Juvenile literature. [1. Physical sciences.] I. Time-Life Books. II.
Series.
Q163 .P576 2000
530—dc21   99-057136

10 9 8 7 6 5 4 3 2 1

OTHER PUBLICATIONS

TIME-LIFE KIDS
Library of First Questions and
  Answers
A Child's First Library of Learning
I Love Math
Nature Company Discoveries
Understanding Science & Nature

HISTORY
Our American Century
World War II
What Life Was Like
The American Story
Voices of the Civil War
The American Indians
Lost Civilizations
Mysteries of the Unknown

Time Frame
The Civil War
Cultural Atlas
SCIENCE/NATURE
Voyage Through the Universe

DO IT YOURSELF
Custom Woodworking
Golf Digest Total Golf
How to Fix It
The Time-Life Complete Gardener
Home Repair and Improvement
The Art of Woodworking

COOKING
Weight Watchers® Smart Choice
  Recipe Collection
Great Taste-Low Fat

For information on and a full description of any of the Time-Life Books series
listed above, please call 1-800-621-7026 or write:

Reader Information
Time-Life Customer Service
P.O. Box C-32068
Richmond, Virginia 23261-2068